Calibration

ISA Technician Series

Calibration: A Technician's Guide, second edition, is part of a series of books covering the most common work functions performed by today's automation and control systems technicians. The Technician Series includes these books:

- *Calibration: A Technician's Guide*
- *Loop Checking: A Technician's Guide*
- *Project Management: A Technician's Guide*
- *Start-Up: A Technician's Guide*
- *Troubleshooting: A Technician's Guide*

More information on the series is available at www.isa.org/products/technician-series-1.

Calibration

A Technician's Guide

Second Edition

Michael P. Cable
Alan W. Bryant

ISA Technician Series

International Society of Automation
Setting the Standard for Automation®

WILEY

Published by John Wiley & Sons, Inc., Hoboken, New Jersey.

Library of Congress Cataloging-in-Publication Data

Names: Cable, Mike P. (Industrial technician) author | Bryant, Alan W., 1967- author
Title: Calibration: A Technician's Guide / Michael P. Cable, Alan W. Bryant.
Description: Second Edition. | Hoboken, New Jersey : Wiley, [2026] | Series: ISA Technician Series | Includes bibliographical references and index.
Identifiers: LCCN 2026004894 (print) | LCCN 2026004895 (ebook) | ISBN 9781394402410 Hardback | ISBN 9781394402434 adobe pdf | ISBN 9781394402427 epub
Subjects: LCSH: Engineering instruments–Calibration
Classification: LCC TA165 .C12 2026 (print) | LCC TA165 (ebook)
LC record available at https://lccn.loc.gov/2026004894
LC ebook record available at https://lccn.loc.gov/2026004895

Cover Design: Wiley
Cover Image: © NeoLeo/stock.adobe.com

Set in 9.5/12.5pt STIXTwoText by Straive, Chennai, India

Contents

List of Figures

List of Tables

About the Book

The dates and editions for standards and other references in this chapter are current as of publication. The reader is encouraged to check for later revisions or editions.

About the Authors

Michael P. Cable is a Level 3 Certified Control System Technician (CCST) and certified Project Management Professional (PMP). He is currently the Lead Controls Engineer at United Therapeutics in Durham, North Carolina.

Cable started his career as an electronics technician in the Navy Nuclear Power Program serving as a reactor operator and engineering watch supervisor aboard the USS *Los Angeles* submarine and then at the A1W nuclear prototype in Idaho Falls, Idaho. After leaving the Navy, he started his civilian career at Performance Solutions, performing technical services for the pharmaceutical industry. His 11 years there were highlighted by an assignment to Eli Lilly Corporate Process Automation, where he managed instrument qualification projects, and then he started a Calibration Services division within Performance Solutions. Cable has since worked at cell therapy companies in Operations Technology, implementing various electronic systems (MODA-EM, MODA-ES, FactoryTalk PharmaSuite, MasterControl, ProCal, SAP, and Sage). He has also been a consultant with PCI in Raleigh, helping companies bring value-added improvements to their calibration programs and their implementations of Blue Mountain Regulatory Asset Manager.

Alan W. Bryant is a licensed professional engineer in Texas with a more than 30-year career in the upstream oil and gas industry. Early in his career, he spent time in the field working with instrumentation, gaining hands-on experience that laid the foundation for his experience as a control system engineer. Over the years, Bryant has successfully led automation and multidiscipline projects, bridging technical knowledge with practical application. A longtime volunteer with the International Society of Automation (ISA), he is dedicated to supporting the growth of control system professionals. This book on calibration leverages Bryant's field experience and technical insights to provide control system technicians with practical, effective guidance for their work.

Acknowledgments

I'd like to thank Tom Stevens for running a company in which people are allowed to pursue their interests and, more importantly, because of the support and encouragement he has given me over the years. I'd also like to thank those professionals that I have worked with throughout my career who have contributed in so many ways to my development and the development of others throughout the industry. I need to thank those who helped me early on with performing calibrations in an industrial environment: Bill and Randy at the Eli Lilly (Greenfield) Instrument Shop. Also, thanks to my last supervisor in the Navy, Master Chief Lane Phillips, who taught me many things technically but, most importantly, he taught me the importance of character. I also appreciate Liegh Elrod at ISA, who has been diligent in finding and working with other authors to update this book. Thanks to Alan Bryant for taking on and doing a great job with the second edition.

–Michael P. Cable

I am grateful to my wife, Rhonda, for her patience and support, allowing me to dedicate countless weekends to this project. Thank you to Leo Staples, James Haw, and Michael Williams for their reviews and constructive feedback, which greatly shaped the direction and content. I am also thankful to Liegh Elrod at ISA for giving me the opportunity to undertake this work, providing steadfast support throughout the process, and patiently waiting for me to deliver. Additionally, I am grateful for the support and encouragement I received from the talented individuals at Emerson, Endress+Hauser, Fluke, Beamex, and MSA. Your expertise and assistance have been instrumental in completing this work. Thank you all for your contributions and belief in this endeavor.

–Alan W. Bryant

Introduction

The development of the Certified Control System Technician (CCST) certification process began by identifying the knowledge, skills, and abilities needed for competent job performance. The International Society of Automation (ISA) worked with control systems technicians and supervisors to identify critical job components through interviews, surveys, observation, and group discussions. The knowledge and skill basis for the questions on the current versions of the tests was derived from the actual practice of control systems technicians and developed based on a job task analysis study conducted in 2019.

The 2019 version of the CCST groups tasks into four domains (see Table I.1). These domains align closely with current industry practices and with the skills control systems technicians need in modern industrial settings. Table I.1 lists the typical percentage of questions on the CCST exam for each domain.

The highest percentage of questions on the CCST Level I and Level II exams is from Domain 1 (Calibration, Maintenance, Repair, and Troubleshooting). This book focuses on calibration-related tasks from all the domains, but primarily Domain 1. These tasks are listed in Table I.2 [1].

There are three levels of certification:

1) Level I: CCST
2) Level II: CCST Specialist
3) Level III: CCST Master

This book is intended to serve as a

- study guide for the calibration tasks in the CCST certification levels I, II, and III;
- reference for technicians who perform process instrument calibrations; and
- reference in the classroom for students pursuing studies related to instrumentation.

This text is intended for Level I, II, and III certified control system technicians (CCSTs) who perform calibration tasks on instrumentation during the course of a facility's operations, as well as noncertified control system technicians who aspire to hone their calibration skills. Level III technicians might also be involved in planning a calibration program and creating calibration procedures. Level I and II technicians might be interested in the Level III material, but they will not be

Table I.1 CCST domains, 2019.

Test Domain	Level I Technician	Level II Specialist	Level III Master
1. Calibration, Maintenance, Repair, and Troubleshooting	75%	64%	20%
2. Project Start-Up, Commissioning, Loop Check, Project Organization, and Planning	15%	19%	14%
3. Documentation	10%	11%	17%
4. Administration, Supervision, and Management	0%	6%	49%

Adapted from the ISA CCST Exam Content Comparison Document for 2019 Exam Update. © International Society of Automation. Used with permission of ISA.

2.2 Planning the Program

CCST Level 3

Figure I.1 Example header for Level 3 content.

expected to know it. Sections intended for Level III technicians are identified in the text as shown in Figure I.1.

The calibration examples presented in this text are for typical process industry measurements, such as pressure, temperature, level, flow, and gas detection, but the calibration principles will apply to any industrial measurements, such as those in building automation, electric power distribution systems, and manufacturing. The reader should focus on understanding the principles and not on the details of a specific type of device.

Many of the other tasks in the four CCST domains are covered in the other books in this series available from ISA:

- *Loop Checking: A Technician's Guide* by Harley M. Jeffery and Paul M. Franklin
- *Troubleshooting: A Technician's Guide* by William L. Mostia Jr.
- *Start-Up: A Technician's Guide* by Diane R. Barkin
- *Project Management: A Technician's Guide* by Leo Staples

ISA's CCST study guides provide example problems similar to the CCST exam questions. Reviewing the study guides will help students understand the topics that will be covered on the exam.

- *CCST Study Guide Level I*
- *CCST Study Guide Level II*
- *CCST Study Guide Level III*

Table I.2 CCST tasks covered in this book.

Code	Task	Level I	Level II	Level III
0101	Calibrate a device using appropriate test standards, recommended procedures, and manufacturer's specifications on instrument data sheets to record as-found readings, evaluate as-found readings against specified tolerance, make calibration adjustments as required, and record as-left data.	X	X	X
0102	Assess the condition of a device through documentation review, inspection, and testing in order to make adjustments to maintain device performance and accuracy and to determine the need for repairs.	X	X	X
0105	Obtain applicable documents required to perform device calibration, testing, troubleshooting, and maintenance, and review the documents for accuracy and completeness.	X	X	
0107	Review the SDSs, appropriate PPE, and safety procedures to prepare for removing and decontaminating a device from operational service.	X	X	
0108	To calibrate the device, select the correct testing equipment using existing documentation, appropriate calibration procedures, and visual inspection.	X	X	
0109	Use documentation and field inspection to verify that appropriate utilities and equipment are available and operational so devices can be tested safely and effectively.	X	X	
0110	To maintain safety in hazardous environments, apply the proper fittings, terminations, and electrical barriers for instruments and devices by utilizing the correct procedures according to local regulatory codes and standards.	X	X	
0111	Use appropriate testing equipment to measure and detect electrical and/or process values to ensure proper loop installation and performance.	X	X	
0114	Isolate a process component from an operational system by following maintenance documentation and appropriate safety and operating procedures in order to perform maintenance functions.	X	X	
0121	To avoid bodily harm or death, wear the appropriate PPE required for each hazard risk category, consistent with arc flash ratings.	X	X	X
0202	To ensure safe and efficient device operation, visually inspect the loop components to verify that they are installed correctly according to the applicable documentation.	X	X	
0301	Provide data, reports, and technical support for regulatory agencies to satisfy regulatory compliance requirements.		X	X
0303	Document calibration, maintenance, troubleshooting, and repair using appropriate forms, calibration labels, and/or electronic records to provide a permanent record of changes and device history.	X	X	

(Continued)

Table I.2 (Continued)

Code	Task	Level I	Level II	Level III
0401	Identify electrical classification or other hazards in an area, and determine appropriate procedures to be followed for safe and effective operation in that area.		X	X
0402	Administer/coordinate work orders to facilitate proper calibration, repair, and maintenance.			X
0403	Coordinate the removal and decontamination of a device from operational service by reviewing the SDSs, the appropriate PPE, and safety procedures.		X	X
0404	Coordinate with all affected personnel to isolate a process component from an operational system so that testing or maintenance can be performed.			X
0407	Develop procedures for using simulation and test equipment to verify proper loop functionality, calibration, and maintenance.			X
0413	Update and maintain certified testing equipment, documentation, recommended procedures, and out-of-tolerance reporting to facilitate the operation of instruments and devices necessary for safety and quality assurance.			X

PPE, personal protective equipment; SDSs, Safety Data Sheets.
Adapted from the ISA CCST Exam Content Comparison Document for 2019 Exam Update. © International Society of Automation. Used with permission of ISA.

Reference

1 International Society of Automation (ISA), 2019, CCST Level I, II, III Examination Content Outline, ISA. https://www.isa.org/getmedia/064262e2-80ab-4f36-970c-6abd72058754/CCST-Level-I-II-III-Blueprint-Comparison-Final-20180430.pdf. Accessed April 6, 2025.

1

Calibration Principles and Terminology

1.1 What Is Calibration?

Is an instrument reading in the control system "right?" How can we find out whether or not it is right? What is the right reading? Is the reading close enough to the right answer? How close is close enough? What do we do if the reading isn't right?

The purpose of a calibration program is to answer these questions and resolve issues methodically. This book introduces the concepts of calibration and a calibration program from the perspective of a control system technician (CST) and discusses how these concepts can be applied to a few of the most common process measurement instruments as examples. It does not provide examples for all instruments used in control systems, but the concepts can be applied to any instrument.

We decide whether a reading is "right" by comparing it to a measurement we know is very close to right. This process is called *calibration.*

> ***calibration:*** procedure *of checking or adjusting* (*by comparison with a reference standard*) *the accuracy of a measuring instrument*
>
> (*International Society of Automation [ISA],* [1], Clause 3.1.3).

Calibration compares a measuring instrument that has an unknown amount of uncertainty against a standard of higher accuracy that has a known amount of uncertainty. The measurement standard gives an answer within a range of accuracy that, for practical purposes, is the "right" answer.

Calibration is used to detect, correlate, adjust, rectify, and document the accuracy of the instrument being calibrated. In general, the steps are:

1) Find the appropriate measurement standard for comparison.
2) Apply an input to the instrument being checked.
3) Compare the instrument reading to the standard reading.
4) If the instrument reading is within the specified tolerance, record the result.
5) If the instrument reading is not within the specified tolerance, follow the steps detailed in the calibration procedure (e.g., repair, replace, adjust, and record).

Note that calibration is not the same as configuration. A smart instrument can be configured for a specific range of measurement, but that process does not compare the instrument reading to a reference standard.

1.2 Calibration-Related Terminology

1.2.1 Calibration Range

The *calibration range* is defined by the upper and lower range values, as shown in Figure 1.1.

- *Upper range limit (URL)* is the upper end of the instrument's capability.
- *Lower range limit (LRL)* is the lower end of the instrument's capability.
- *Instrument range* or *full range* is the difference between the upper and lower range limits.
- *Upper range value (URV)* is the upper end of the configured measurement range.
- *Lower range value (LRV)* is the lower end of the configured measurement range and the minimum output of the transmitter. It is also commonly called the *zero value.*
- *Span* or *calibration range* is the difference between the upper and lower range values.

The calibration range may differ from the instrument range, which refers to the instrument's capability. For example, an electronic pressure transmitter may have a nameplate instrument range of 0 to 750 pounds per square inch gauge (psig) and an output of 4 mA to 20 mA. However, the design calls for the instrument to be calibrated for 0 to 300 psig = 4 mA to 20 mA. Therefore, the calibration range (span) would be specified as 0 to 300 psig = 4 mA to 20 mA. In this example:

LRV (zero) = 0 psig input, 4 mA output
URV = 300 psig input, 20 mA output
Input span = 300 – 0 = 300 psig
Output span = 20 mA – 4 mA = 16 mA

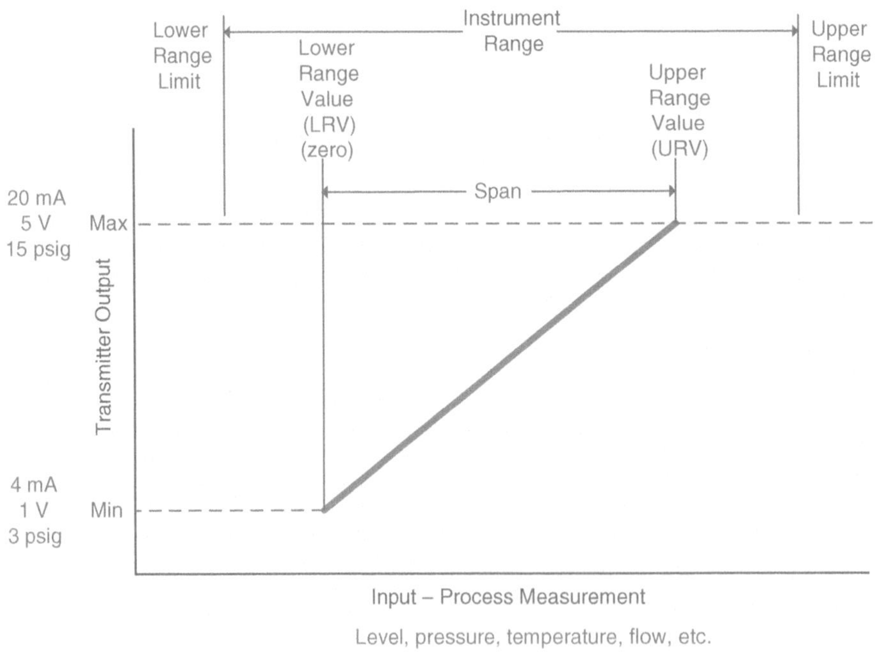

Figure 1.1 Calibration range terminology. Max, maximum; Min, minimum.

With regard to instrument accuracy, when accuracy is stated as a "percentage of full scale," it is important to know if the manufacturer is referring to the instrument range or the calibrated range.

1.2.2 Error and Accuracy

- **Error** – The difference between an indicated value and the actual value. For example, if the instrument reads 90, but the actual value is 100, then the error is 10.
  - **Zero error** – This error is approximately equal over the entire scale.
  - **Span error** – This occurs when the error at full scale differs from the error at zero, often proportional to the input.
  - **Linearity error** – The response is not linear.
  - **Hysteresis error** – The error is different when the input is increasing than when the input is decreasing.
- **Accuracy** – The degree to which an indicated value matches the actual value of the measured variable. The accuracy rating of a device is the limit that errors will not exceed when it is used under specified conditions (i.e., how close to "right" the device can accomplish). Instrument manufacturers typically provide this rating as a percentage of the actual reading or a percentage of full scale (Figure 1.2).

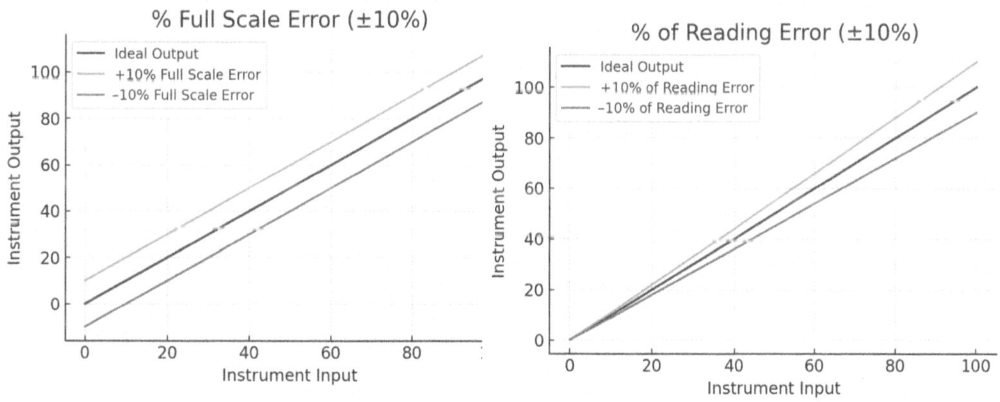

Figure 1.2 Percentage of full scale versus percentage of reading error.

- **Percentage of full scale** – The difference between the indicated value and the actual value, expressed as a percentage of the device's full range of values (LRL to URL). The error would be the same for readings at the bottom of the scale and the top of the scale.
  Example:
  - Device range: 0 to 200
  - Accuracy: ±1% of *full scale*

Reading	Error Range	Actual Value Between	
50	1% of 200 = ±2	48	52
150	1% of 200 = ±2	148	152

- **Percentage of reading** – The difference between the indicated value and the actual value, expressed as a percentage of the actual value. The error would be smaller for readings on the low end of the scale than for readings on the high end.
  Example:
  - Device range: 0 to 200
  - Accuracy ±1% of *reading*

Reading	Error Range	Actual Value Between	
50	1% of 50 = ±0.5	49.5	50.5
150	1% of 150 = ±1.5	148.5	151.5

1.2.3 Precision and Repeatability

Precision represents the repeatability of measurement data. If a measurement is made repeatedly under the same conditions by the same person with the same equipment, the measurement will not always give the same result. Precision (or repeatability) represents how close the repeated measurements are to each other.

It is important to understand the difference between the terms *accuracy* and *precision* because they are often mistakenly used to mean the same thing. Imagine repeatedly trying to hit a target as shown in Figure 1.3.

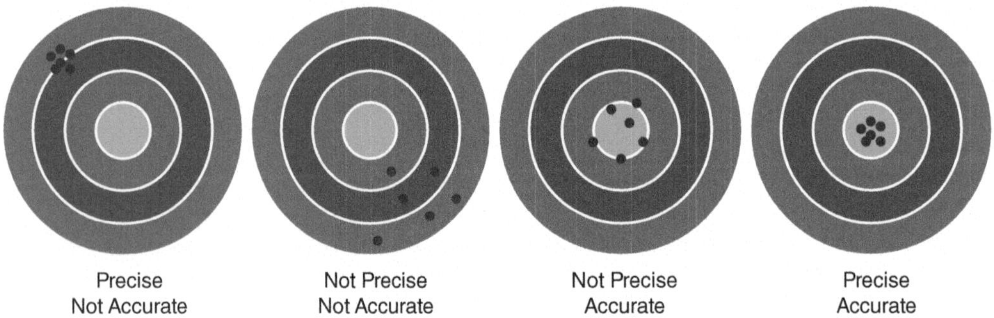

Figure 1.3 Precision versus accuracy.

With respect to calibration, think of the calibration instrument as the target and the dots representing measurements from the device to be calibrated. If the goal is to get the measuring device to give readings inside the inner circle, that circle represents the tolerance.

1.2.4 Tolerance

Tolerance is the permissible limit of variation in a measurement with respect to the actual value. Tolerance is "how close is close enough."

Every calibration should be performed to a specified tolerance. The tolerance will be defined by an organization's calibration program and given in the calibration document. It might be given for each device or a device loop. The tolerance will typically be specified in the measurement units of the reading (e.g., psig, °F, V). By specifying an actual value, mistakes caused by calculating the percentages of span or reading are eliminated. Tolerances should be specified in the units of measurement for the calibration.

1.2.5 Stability

Stability is the maximum change over a time period. The stability of a device is usually given in terms of the maximum expected change in error over a time period, for example, ±0.25% of reading for 5 years. This is how long manufacturers promise that a device will meet the accuracy specifications.

Some manufacturer data sheets also list "zero stability." *Zero stability* is the ability of an instrument to maintain its calibrated zero reading over a specified period of time. For example, it might be given as ±0.1% of URL per year.

1.2.6 Test Uncertainty Ratio

Test uncertainty ratio (TUR) is the relationship between the accuracy of the test standard and the accuracy of the instrument under test. ISA-RP105.00.01-2017 recommends that standards used for calibration should be at least 4 times more accurate than the required tolerance of the equipment being calibrated (International Society of Automation [ISA], [1]). Therefore, the test equipment (such as a field standard) used to calibrate a process instrument should be 4 times more accurate than the accuracy required of the instrument; the laboratory standard used to calibrate the field standard should be 4 times more accurate than the field standard, and so on. This is a TUR of 4:1.

If multiple calibration equipment components are used to calibrate a loop/component, the aggregate uncertainty of the calibration equipment components should be calculated using the root sum squared (RSS) method, and that aggregate uncertainty should be 4 times greater than the loop/device tolerance ([1], Clause 5.7).

For example, if a loop has been assigned a tolerance of ±1 psig, the calibration standard(s) must have an overall uncertainty of ±0.25 psig.

1.2.7 Uncertainty

A calibration technician will sometimes come across these terms in product literature:

- Uncertainty
- Repeatability
- Probability
- Confidence interval

Uncertainty is a lack of knowledge of the true value of a variable. For calibration equipment, uncertainty is given by the *probability* that a measurement will fall within a certain range. The range is called the *confidence interval*. A 95% confidence interval means that a reading will fall within the given range 95% of the time.

The terms *repeatability*, *precision*, and *confidence interval* are often seen in analytical equipment specifications.

1.2.8 Combined Effects on Uncertainty: Worst-Case Versus Probable

Calibration technicians (especially Level III certified control system technicians [CCSTs]) should be aware of basic uncertainty factors, such as environmental effects, and how to combine multiple calibration equipment accuracies to arrive at a single calibration equipment accuracy. Combining multiple calibration equipment or process instrument accuracies is done by calculating the square root of the sum of the squares.

A device's overall uncertainty is usually a combination of the stated accuracy in addition to influences of factors such as:

- Ambient conditions (temperature, humidity, etc.)
- Process conditions (process pressure, temperature)
- Supply voltage
- Stability (maximum change over a time period)

These errors would be combined for a device to find the total probable error (TPE) for that device (Equation 1.1). To perform this calculation correctly, all errors should be expressed relative to the actual loop/component application measurement span and in the same manner (e.g., percentage of full scale, percentage of reading, or units of measure) ([1], Clause 5.5).

$$\text{TPE} = \sqrt{\text{accuracy}^2 + \text{error1}^2 + \dots \text{error2}^2} \tag{1.1}$$

The TPE represents the highest error that is likely to be seen for practical purposes. This assumes that not everything is at its worst error at the same time.

The *worst-case error* (WCE) is the sum of all the error components (Equation 1.2):

$$\text{WCE} = \text{accuracy} + \text{error1} + \dots \text{error2} \tag{1.2}$$

WCE is the worst error that could occur if everything that could go wrong did go wrong all at once for the device.

The same RSS calculation that is given here for a device is also used to calculate the TPE for the combination of all devices in a loop or for combining the uncertainties of multiple calibration equipment.

1.2.9 Zero and Span Errors

Zero error means that the instrument does not read or output the correct value when the input is at the minimum value. This shifts the entire calibration line up or down by a constant amount (Figure 1.4).

Table 1.1 is an example of a pressure transmitter scaled 0–100 psig = 4–20 mA with a zero error of 1.25 psig.

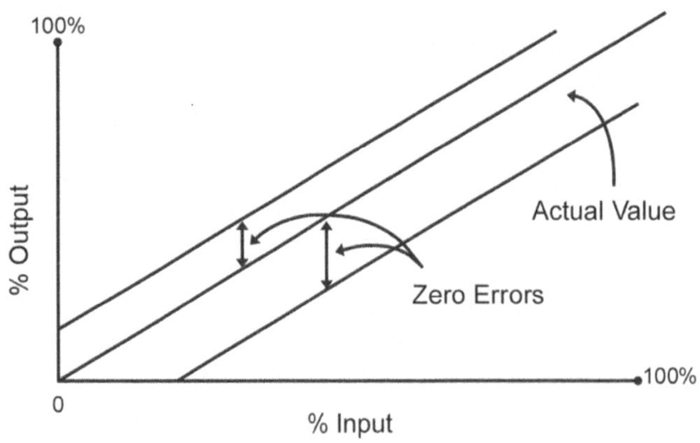

Figure 1.4 Zero error.

Table 1.1 Zero error example.

Input	Expected Output	Actual Output
0 psig	4 mA (0 psig)	4.2 mA (1.25 psig)
50 psig	12 mA (50 psig)	12.2 mA (51.25 psig)

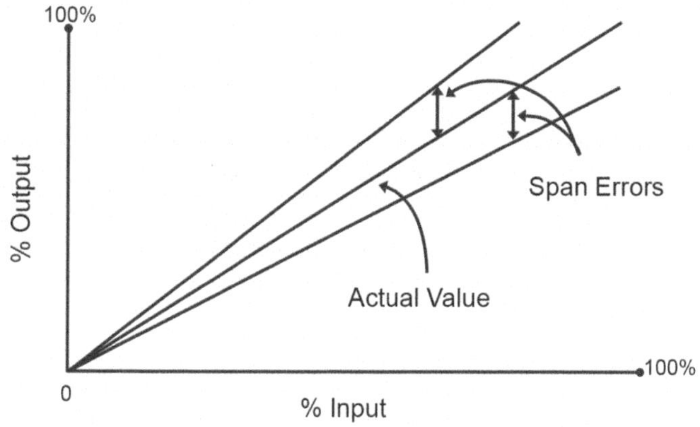

Figure 1.5 Span error.

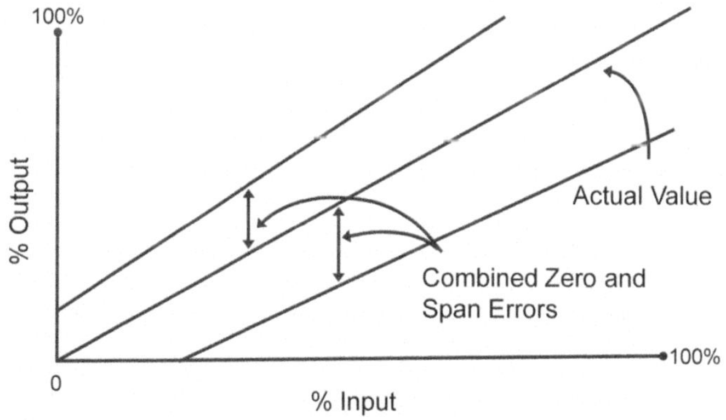

Figure 1.6 Combined zero and span errors.

Span error exists when the error at full scale differs from the error at zero pressure. This error is often proportional to the input, so it affects the slope of the calibration line (Figure 1.5).

If a device has both zero and span errors, they combine to shift the calibration line up or down and also change the slope of the line (Figure 1.6). Zero and span errors can often be corrected by adjusting the transmitter configuration.

1.2.10 Linearity and Hysteresis Errors

Typically, an instrument's calibration is checked at several points throughout its calibration range because sometimes the error is not *linear*; that is, the error varies along the range of an instrument, as illustrated in Figure 1.7.

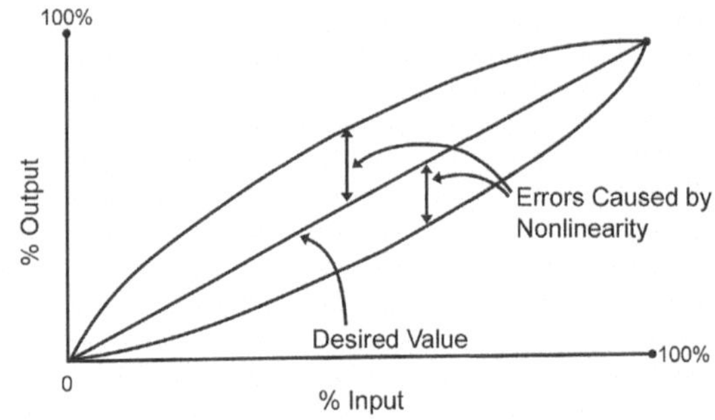

Figure 1.7 Nonlinear error example.

Hysteresis is the measure of the difference in the response of a device as the input signal increases from a minimum to a maximum value and subsequently decreases from maximum to minimum value over the same range, as shown in Figure 1.8. For continuous measurement, the instrument's output is different when the input is going up than when the input is going down. The switch with hysteresis will transition at a different value when the input is going up than when the input is coming down.

Hysteresis can be troublesome on process indicators, sensors, and transmitters. However, a hysteresis feature is built into some process switches to prevent the switch from toggling on and off when the process variable is at the set point. Think of a thermostat in a house that is set at 72 °F. It might come on at 73 °F and turn off at 71 °F.

Hysteresis is expressed in percentage of full scale (%FS).

Instruments with mechanical movements, such as pressure gauges, valve positioners, and current-to-pneumatic (I/P) transducers, develop hysteresis. This is often caused by friction and worn mechanical components. There is no way to adjust for hysteresis during calibration.

Hysteresis can be evaluated by obtaining calibration data with the input applied in both the increasing and decreasing directions, as illustrated in Table 1.2 for a device with a full scale of 100 psig.

When evaluating hysteresis, it is important to approach the target value without overshooting the applied input. If the input value exceeds the increasing target value, decrease the applied input to the previous test point and retry. If the input value passes the decreasing target value, increase the applied input above the previous test point and repeat the test.

Table 1.2 Example of hysteresis error.

Applied Input (psig)	As-Found Values (↑) (psig)	As-Found Error (↑) (psig)	As-Found Values (↓) (psig)	As-Found Error (↓) (psig)	Hysteresis (%FS)
10	9	−1	10	0	1%
25	24	−1	25	0	1%
50	49	−1	51	1	2%
75	74	−1	76	1	2%
90	90	0	91	1	1%

%FS, percentage of full scale.

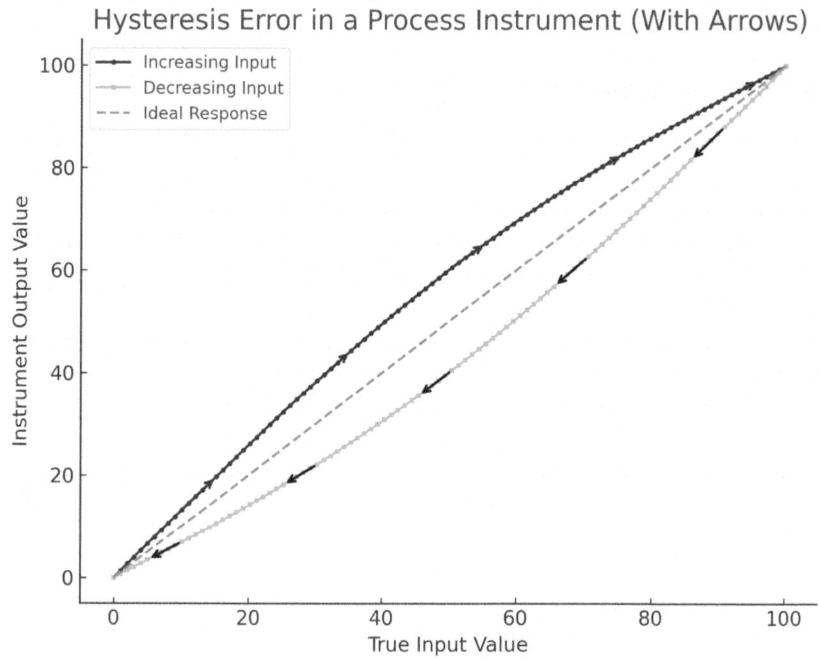

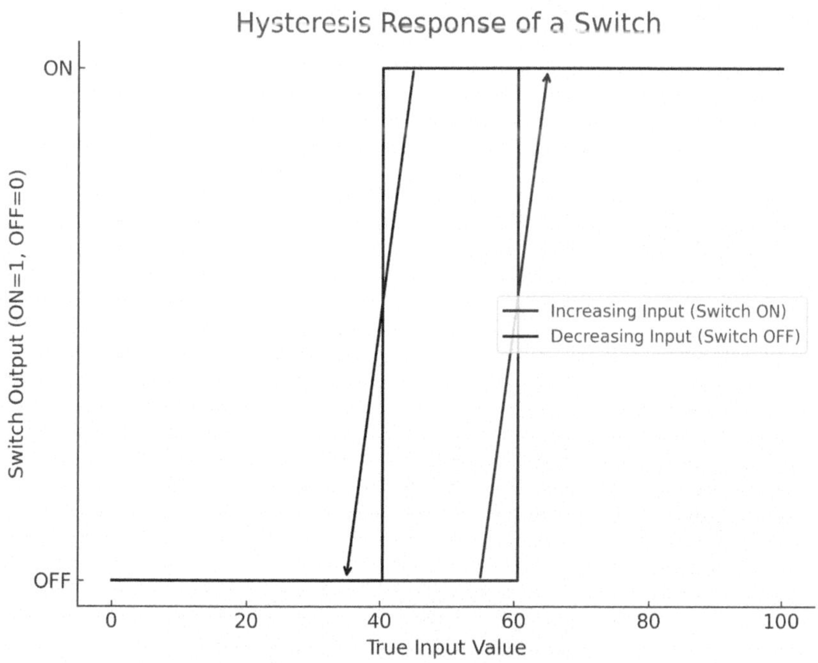

Figure 1.8 Transmitter and switch hysteresis examples.

1.3 Calibration Methods

1.3.1 Loop Calibration Versus Device Calibration

Loop calibration is performed from the sensor to all loop indications with all the loop components connected, as shown in Figure 1.9. This is the recommended best practice from ISA-RP105.00.01-2017 [1]. This concept is important because it checks the performance of all components working together to measure a process parameter (e.g., sensor, transmitter, and input card). Alarms, controls, and protective functions within the control or safety system might also be verified during a loop calibration.

A source (e.g., pressure or temperature) is applied to the sensor and adjusted to various points through the calibration range. The loop indication (e.g., the operator interface reading) is recorded. The sensor and transmitter outputs might also be recorded. The goal is to determine whether the loop provides a measurement within the stated tolerance.

If the loop is not within tolerance, troubleshooting may be required. The company's procedures for troubleshooting and adjustments should be followed, which might include calibrating individual devices.

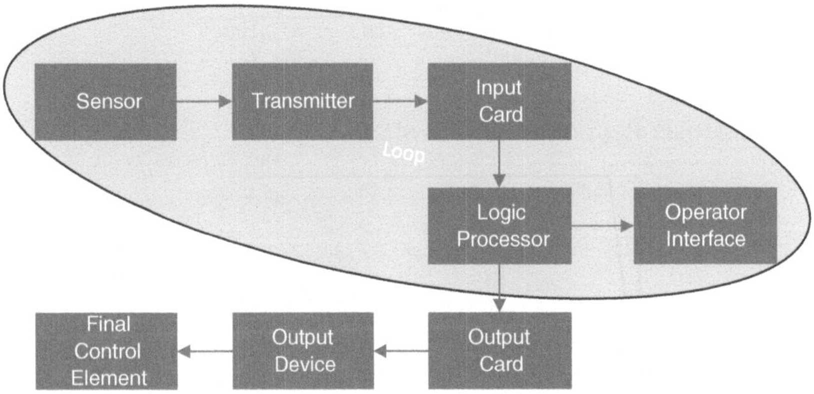

Figure 1.9 Loop calibration.

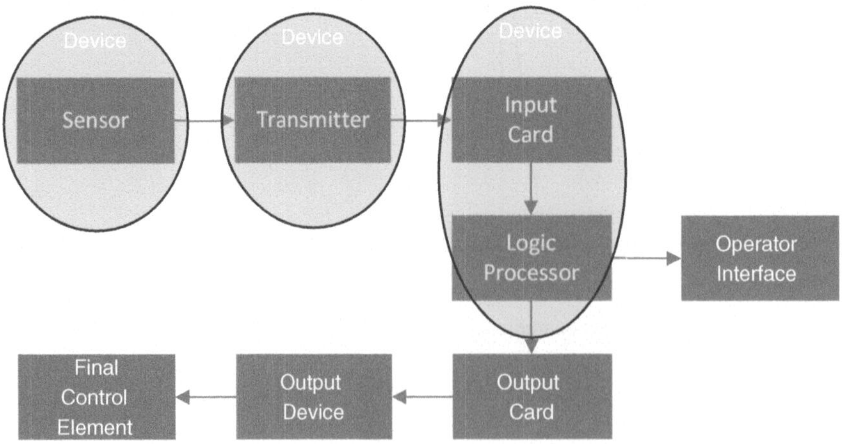

Figure 1.10 Device calibration.

Device calibration is performed only on one device in a loop (Figure 1.10). A known source is applied to the input, and the output is measured at data points throughout the calibration range. If necessary, the instrument is adjusted, and calibration is checked again.

New devices are usually calibrated before installation to confirm that they work correctly. If a loop calibration fails to meet the required tolerance, individual devices in the loop might need to be calibrated to find the cause. Device calibration might also be used when safety or operation considerations prevent testing devices or loops in a running system.

Expert Advice

Many companies still individually calibrate each device in a system and pay little attention to the combined effect each device plays in the overall accuracy of the loop. The reason most frequently given by those who oppose loop calibrations is the idea of "de-calibrating" a transmitter to make the reading correct on the data acquisition system, control system, or final indicator. In reality, no sensor is perfect, and the output of some element in the loop is required to bring the loop into calibration.

An example of a measurement error that occurs when loop calibration is not performed is the process of calibrating a temperature device without the sensor in the loop. First, the technician disconnects the temperature element from the device (transmitter or measurement system). Next, the technician connects a temperature simulator to the device to be calibrated. This reduces the potential thermal shock to the temperature element and the time to calibrate the temperature device. However, this process ignores the measurement error of the temperature element. How much of a measurement error is introduced depends on many factors. For a Class B Industrial resistance temperature detector (RTD) operating at 500 °C (932 °F), the accuracy is ±2.8 °C or ±5.04 °F. This approach to calibration does not account for the wiring configuration (two, three, or four-wire), self-heating, or the stability of the RTD. Depending on the process, a measurement error of that magnitude can be costly [2].

1.3.2 Bench Versus Field Calibration

A calibration procedure must also indicate whether the calibration is performed in the field or the instrument is uninstalled and calibrated in a shop.

A *bench calibration* (or bench test) is performed in the shop with power supplied from an external source if required. Bench tests are typically performed when new instruments are received, before installation. This provides assurance that the instrument is received in good working condition. This also allows for configuration and calibration in a more favorable environment.

Some periodic calibrations might be performed on the bench when there is no practical way to test the device in place. In this case, the process device is removed from service, disconnected, and taken to the shop for calibration. A calibrated spare might be installed in its place to minimize the process downtime. For example, a critical flow sensor might be sent to a certified flow calibration facility, and a calibrated flow sensor would be installed in its place.

Field calibrations are performed in situ, or in place. The instrument being calibrated is not removed from the installed location. Loop calibration is always done in the field.

Field calibrations may be performed after installation to ensure proper connections and configuration. Periodic calibrations are more likely to be performed in the field as part of a loop calibration. Field calibrations are performed in the environment in which the instrument operates. If the instrument is installed in a harsh environment, it is calibrated for that environment. If the instrument is removed for a bench test and returned to service, some error may be introduced because of the ambient conditions, process conditions, power supply, and orientation.

1.4 Fundamentals of Metrological Traceability

Metrology is the scientific study of measurement. Calibration programs, tools, and methods are rooted in this science. (It is different from meteorology, which is the science of weather forecasting.)

Traceability is the property of a measurement result whereby the result can be related to a reference through a documented, unbroken chain of calibrations, each contributing to the measurement uncertainty [3].

Traceability is accomplished by ensuring the test standards are routinely calibrated against higher-level reference standards. Typically, the standards used locally are sent out periodically to a calibration lab that has more accurate test equipment. The standards from the calibration lab are checked regularly for calibration by higher-level standards, and so on until, eventually, the standards are tested against primary standards maintained to an internationally recognized standard. The higher you go on the calibration pyramid (Figure 1.11), the smaller the uncertainty.

All calibrations should be traceable to a nationally or internationally recognized standard. For example, in the United States, the National Institute of Standards and Technology (NIST) maintains nationally recognized standards and certifies calibration labs. This does not mean a calibration shop must have its standards calibrated by a national laboratory. It means that the calibrations performed are traceable to the national standard through all the standards used to calibrate the standards, no matter how many levels exist between the shop and the national standard (Figure 1.12).

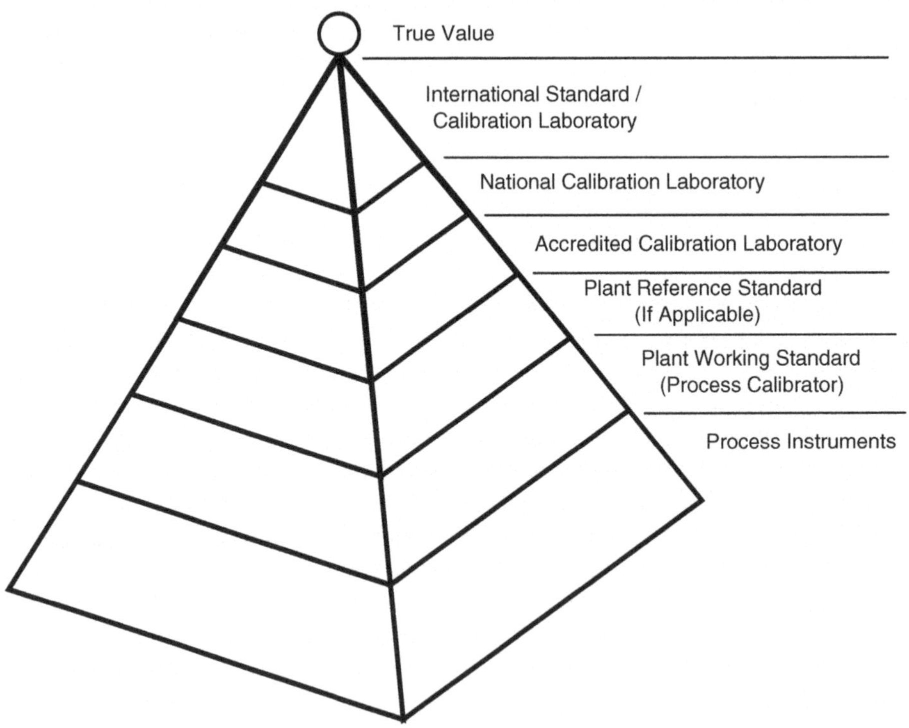

Figure 1.11 Calibration pyramid.

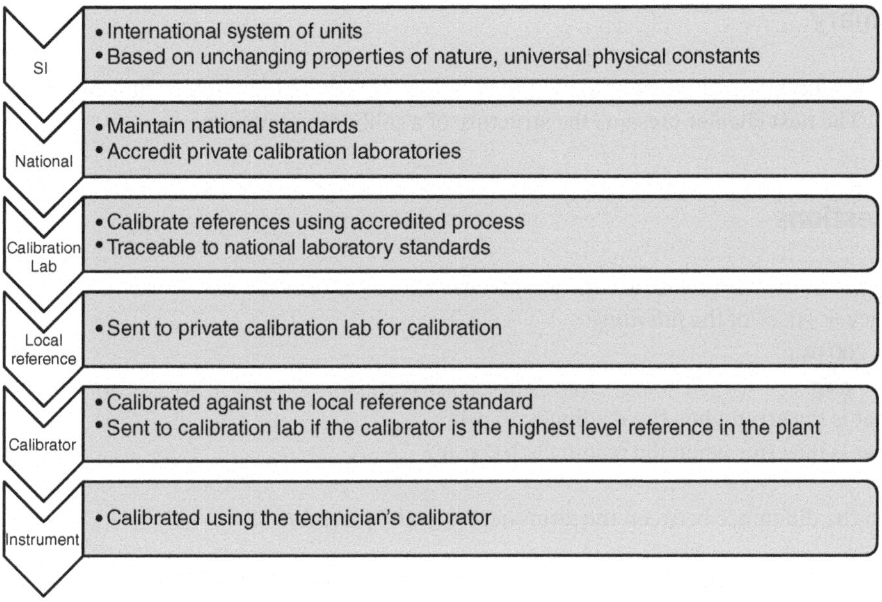

Figure 1.12 Traceability.

The Last Artifact

(This is not on the CCST exam; it is just neat to know. This author thinks the history of metrology over the centuries is fascinating.)

In 2018, all reference standards were redefined by the International Bureau of Weights and Measures (BIPM). The references are now defined in terms of fundamental, unchanging properties of nature and can be reproduced by any organization.

For instance, the reference for mass is now based on a combination of physical constants rather than a piece of metal in a vault in Paris. The International Prototype Kilogram, aka "Le Grande K," was officially replaced by the 2018 vote. The platinum cylinder served for 130 years as the international standard for mass. It resided in international territory just outside Paris in a triple-locked underground vault and was the last artifact of metrology to be replaced by the new system.

BIPM's brochure "The International System of Units" is a good reference. It can be downloaded from their website.

An organization's local calibration standards and field calibration devices must be recalibrated on an interval defined in the organization's calibration program. This interval is generally determined based on the device manufacturer's recommendation plus experience (see Section 2.7).

Of course, detailed recordkeeping is necessary to prove traceability. The calibration technician's role in maintaining traceability is to ensure that the test equipment is within its calibration interval and that the unique identifier is recorded on the applicable calibration data sheet when the instrument calibration is performed. Additionally, when test standards are calibrated, the calibration documentation must be reviewed for accuracy and to ensure that a calibration was performed using traceable equipment. Level III technicians will probably have the responsibility for incorporating traceability and recordkeeping into the calibration program.

1.5 Summary

This chapter introduces calibration definitions and basic concepts that calibration technicians need to understand. The next chapter presents the structure of a calibration program.

Review Questions

1.1 A pressure transmitter's base accuracy is given:
Accuracy = ±0.5% of the *full range*
URL = 200 psig
LRL = 0 psig
a) What is the error when the reading is 50 psig?__________
b) What is the error when the reading is 100 psig?__________

1.2 Explain the difference between the terms *accuracy* and *precision.*

1.3 The term *tolerance* means "how close is close enough" or "how good does it need to be?" Is it better to express tolerance in percentage of reading or in actual units of measure?

1.4 A temperature transmitter is capable of measuring −100 °F to 100 °F, but it is configured to measure 0 °F to 50 °F. Calculate the values for:
URL __________
LRL __________
URV __________
LRV __________
Range __________
Span __________

1.5 A temperature transmitter with a configuration range of 0 °F to 100 °F is found to have the outputs shown in the following table:

Input	Expected Output	Actual Output
0 °F	4.0 mA	4.0 mA
25 °F	8.0 mA	8.1 mA
50 °F	12.0 mA	12.2 mA
75 °F	16.0 mA	16.3 mA
100 °F	20.0 mA	20.4 mA

This is an example of ______________.
a) Zero error
b) Span error
c) Both errors combined
d) Hysteresis

1.6 A temperature gauge with a range of 0 °F to 100 °F is found to have these readings as the temperature is increased and then decreased:

Input	Reading	Error
0 °F	0 °F	0
25 °F	25.1 °F	+0.1
50 °F	50.3 °F	+0.3
75 °F	75.1 °F	+0.1
100 °F	100 °F	0
75 °F	74.9 °F	−0.1
50 °F	49.7 °F	−0.3
25 °F	24.9 °F	−0.1
0 °F	0 °F	0

This is an example of ________________.

a) Zero error
b) Span error
c) Both errors combined
d) Hysteresis

1.7 ISA-RP105.00.01-2017 says the recommended best practice for performing calibrations is:

a) Calibrate just the instrument (device calibration).
b) Calibrate the entire instrument loop, including the instrument to the control system input, to the human-machine interface (HMI; loop calibration).

1.8 Which of the following best describes the term *traceability* of reference standards used for calibration?

a) The ability to calibrate a device using any available reference standard
b) The capability of a field calibrator to store historical calibration data

c) The ability to track a calibrator's accuracy to a known standard following a clear, documented path of calibrations
d) The use of manufacturer specifications to determine the total probable error (TPE) of a measurement

1.9 Regarding traceability, the calibration technician should do one or more of the following. Select **all** that apply.
a) Ensure the test equipment is within its calibration interval before using it to perform calibrations.
b) Send test equipment to a certified lab before every use.
c) Record the equipment's unique identifier on the calibration record sheet.
d) Perform bench calibrations on every instrument.

CCST Level III

1.10 A pressure transmitter's specifications include:
Range = 0 psig to 100 psig
Base accuracy: ±1% of range
Temperature effects: ± 0.5% of range
a) Use the RSS equation to calculate the TPE for the device.
b) Calculate the WCE for the device.

References

Note: The dates and editions for standards and other references in this chapter are current as of publication. The reader is encouraged to check for later revisions or editions.

1 International Society of Automation (ISA), ISA-RP105.00.01-2017, *Management of a Calibration Program for Industrial Automation and Control Systems*. Research Triangle Park, NC: ISA, 2017.
2 Staples, Leo, "A Standardized Approach to Calibration of Automation Instrumentation and Systems," paper presented at Automation Week, October 2011, Mobile, AL.
3 National Institute of Standards and Technology (NIST), 2024, June 17, "NIST Policy on Metrological Traceability." NIST. https://www.nist.gov/calibrations/traceability. Accessed December 17, 2024.

2

The Calibration Program

Much of this chapter is especially relevant to the Level III certified control system technician (CCST) who is involved in planning a calibration program and creating procedures and forms. It should also be interesting to the Level I and II technicians to understand the whys behind the program.

2.1 Introduction

It makes sense that calibration is required to ensure a new instrument provides an accurate indication or output signal when it is installed. But why can the instrument not be left alone if it is operating properly and continues to provide the expected indication?

Instrument error can occur due to a variety of factors: drift, environment, electrical supply, the addition of components to the output loop, process changes, and so on. Periodic calibrations are performed to detect and correct instrument errors. Even if a periodic calibration reveals that the instrument is perfect and no adjustment is required, that would not be known without performing the calibration. Periodic calibrations to specified tolerances using approved procedures are important elements of any quality system.

Accurate, reliable, and repeatable operation of instrument loops is necessary for a facility to operate as intended. This requires planning and implementing a calibration program. The aspects of a calibration program discussed in this book are based on recommended practice ISA-RP105.00.01-2017, *Management of a Calibration Program for Industrial Automation and Control Systems*: "This recommended practice provides guidance on methodologies for calibration of an industrial automation and control system by considering the accuracy of each loop required by the process and, if necessary, adjusting a loop component(s) to achieve the desired loop or component accuracy" ([1], Introduction).

2.2 Planning the Program

CCST Level III

ISA-RP105.00.01-2017 recommends a basic framework for a calibration program consisting of the elements listed in Table 2.1. These elements are covered in this chapter, and related examples are discussed in the following chapters.

Table 2.1 Calibration program elements.

Element	ISA-RP105.00.01-2017 Section
Tagging system	5.2
List of loops or devices	5.2
Criticality rating method and criticality ratings for loops	5.3
Tolerance requirement for each loop	5.4, 5.5
Calibration frequency for each loop	5.6
Calibration equipment requirements	5.7
Calibration recordkeeping system	5.8
Calibration procedures	5.9
Staff qualification and training	5.10
Performing and documenting calibrations	6
Program verification method	7

[1] ISA-RP105, Clause 5. © International Society of Automation. Used with permission of ISA.

An instrument technician who will carry out the calibration and documentation activities defined in the organization's calibration program should be familiar with these concepts and the calibration program. An experienced technician might be involved in planning and managing the program (ISA CCST Level III).

2.3 What to Calibrate? Loops and Components

CCST Level III

ISA-RP105.00.01-2017 says the first steps in developing a calibration program are establishing a tagging system for the instruments/loops and developing a comprehensive list of instruments/loops requiring periodic calibration [1].

Instruments and loops are assigned tags for reference. A tag is used to identify an instrument on all documents and drawings. It is usually also affixed somehow to the physical device in the field (Figure 2.1).

The tagging system is usually established by the organization or project and is used by the calibration program. If no tagging system is established, ANSI/ISA 5.1-2009, *Instrumentation Symbols and Identification* [2], should be used to create a useful tagging system.

First, establish the list of loops to be calibrated using the organization's control system documentation. The documentation needed to plan a calibration program includes the following:

- Instrument index or database with device and loop tags per ANSI/ISA 5.1-2009
- Loop diagrams (ISA 5.4-1991; [3]) or other wiring drawings
- Piping and instrumentation diagrams (P&IDs)

Figure 2.1 Tagged instrument. Adobe Stock Photos.

- Instrument location plans
- Area classification drawings
- Instrument data sheets
- Manufacturer's documentation

See Chapter 3 for more information about each of these. Another ISA text, *Instrumentation and Control Systems Documentation* [4], details the documentation mentioned in this chapter and more.

2.4 How Important Is It? Instrument Criticality Classification

CCST Level III

It is useful to classify instruments in a way that indicates the instruments' importance. An undetected error in an instrument may result in undesirable operation that affects product quality, safety, inventory, and other aspects of production. Additionally, some instruments have calibration requirements specified by contracts or outside agencies.

For these reasons, ISA-RP105.00.01-2017 recommends assigning a criticality classification to each instrument [1]. If the organization does not have a systematic way to determine equipment criticality, a method should be specified for the instruments in the calibration program. ISA-TR91.00.02-2003, *Criticality Classification Guideline for Instrumentation* [5], is an excellent resource for classifying instrumentation. Examples from TR91.00.02-2003 are shown in Figure 2.2. Considerations for assessing criticality include safety, environmental protection, asset protection, process quality impact, and regulatory and contractual requirements.

Criticality classification is used in multiple aspects of the calibration program, such as:

- Relating requirements to regulations and standards.
- Assigning the calibration confirmation interval.
- Identifying training and maintenance requirements.

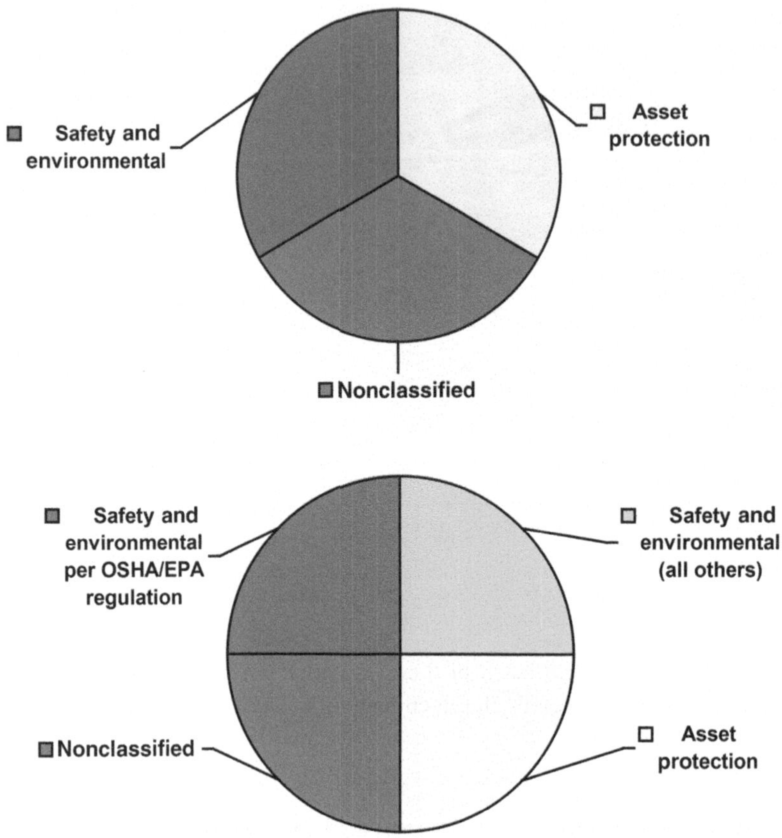

Figure 2.2 Criticality classification examples from ISA-TR91 [5]. EPA, Environmental Protection Agency; OSHA, Occupational and Safety Health Administration. © International Society of Automation. Used with permission of ISA.

2.5 How Good Does It Need to Be? Loop Tolerance

CCST Level III

Tolerance is the permissible limit of variation in a measurement with respect to the actual value; that is, how close to "right" is good enough. The tolerance for each loop will be specified in the calibration procedure. It is not a good idea to set loop tolerance for the company or the site. Performing a criticality assessment should be the first step in determining tolerance requirements for devices and loops.

A technician performing calibrations should be aware of how the loop tolerance is established. Experienced technicians might be involved in establishing the tolerance.

Two methods of establishing loop tolerance are described in ISA-RP105.00.01-2017 [1]:

1) Establish the accuracy needed by the process the loop is controlling, protecting, or monitoring.
2) Establish the combined accuracy that the loop components can theoretically achieve.

Either or both methods can be used; however, the method to be used should be defined in the calibration program. Different methods might be applied to different loops depending on the organization's needs. The methods and associated calculations should be documented in the calibration program.

Theoretical loop accuracy must be calculated in either case. If method 1 is defined for a loop, then method 2 must be calculated to demonstrate that the loop can meet the required tolerance.

Theoretical loop accuracy is calculated using the root sum squared (RSS) method, which combines the rated accuracy of each loop component, such as sensors, transmitters, and input cards (Equation 2.1).

$$\text{tolerance} = \sqrt{\text{device1}^2 + \text{device2}^2 + \ldots \text{device n}^2} \tag{2.1}$$

The RSS method estimates the amount of error that could be reasonably expected, assuming that not all the devices in the loop give the maximum error at the same time. The maximum possible error would be just the sum of the errors of each component, but it is unlikely that all the devices would give the maximum error at the same time, so the RSS method is used.

Each possible source of error for each loop component must be included (e.g., stated accuracy, ambient effects, age, process conditions, and supply voltage variation). All errors must be expressed in the same manner (percentage of reading or percentage of scale) and relative to the same measurement span.

The tolerance specified in measurement units should be used for the calibration requirements performed at the operating facility. Every calibration should be performed to a specified tolerance. Tolerances should be specified in the units measured for the calibration. By specifying an actual value, mistakes caused by calculating the percentages of span or reading are eliminated.

Example 2.1 Calibrate a 0–300 psig pressure transmitter with a specified calibration tolerance of ±2 psig. The tolerance of the 4–20 mA transmitter output would be as shown in Equation 2.2:

$$\frac{2\ \text{psig}}{300\ \text{psig}} \times 16\ \text{mA} = 0.1067\ \text{mA} \tag{2.2}$$

The calculated tolerance is rounded down to 0.10 mA because rounding to 0.11 mA would exceed the calculated tolerance. Both ±2 psig and ±0.10 mA tolerances should appear on the calibration data sheet if the remote indications and output milliamp signal are recorded.

The manufacturer's specified accuracy rating for this instrument may be 0.25% full scale, but the calibration is performed to the application's required tolerance of ±2 psig instead of the manufacturer's specification.

2.6 When Should It Be Calibrated? Calibration Confirmation Interval

CCST Level III

How often should a device or loop be calibrated? The calibration program should define the calibration frequency for each loop. ISA-RP105.00.01-2017 [1] recommends considering the:

- Criticality of the device or loop.
- Performance history of the loop/device.
- Ruggedness and stability of the loop components.
- The operating environment.

Additional considerations include regulatory requirements for checking environmental protection loops, contractual requirements for custody measurement loops, and testing intervals promised in safety integrity level calculations for safety instrumented systems.

Manufacturer's recommendations may be considered when there is no historical performance data to use as a basis.

One method is to use the manufacturer's stated stability for devices as a basis for establishing the calibration interval for a device [6]. This is a five-step process:

1) Determine the required tolerance.
2) Determine the operating conditions.
3) Calculate the total probable error (TPE) for the device in the given service.
4) Get the stability specification from the manufacturer's literature.
5) Calculate the calibration frequency based on how long it takes the TPE to reach the specified tolerance based on stability.

Equation 2.3 is an example of a calculation for determining the calibration interval for a differential transmitter. Here is the data required for the equation:

- Upper range limit (URL) = 1,000 inches of water (in. H_2O)
- Calibrated span = 0 to 250 in. H_2O
- Specified tolerance = ±0.5% of span = ±1.25 in. H_2O
- Calculated TPE at operating conditions = ±0.155% span = ±0.387 in. H_2O
- Stability specification: ±0.15% URL for 15 years = ±0.00833 in. H_2O per month

$$\text{calibration interval} = \frac{(1.25 - 0.387)\ \text{in. H}_2\text{O}}{0.00833\ ^{\text{in. H}_2\text{O}}\!/_{\text{mo.}}} = 104\ \text{mo.} \tag{2.3}$$

Calibration intervals may be adjusted over time. Once several calibrations have been performed, the calibration history of the device may be used to adjust the calibration interval. If the as-found calibration data of a particular instrument does not meet the calibration tolerance, the calibration interval may be shortened. The interval may increase if several instruments with a specific manufacturer/model number are always well within the calibration tolerance. This is called *reliability-centered maintenance.*

"If a high calibration fail rate is observed, before changing the calibration interval, check for correct application and specifications of the device, inappropriate test equipment, unqualified calibration personnel, changes to location environment, or other possible causes." ISA RP105.00.01-2017 ([1], Clause 5.6).

Level III technicians might be involved with establishing and updating calibration frequencies. The calibration frequency for each loop will be documented.

Some form of recordkeeping system must exist to support the program. This might be the organization's computerized maintenance management system (CMMS) or a system dedicated to the instrumentation. A CMMS is mostly for scheduling and work-order management, such as scheduling periodic calibration and issuing calibration work orders. The CMMS may have an optional module to store as-found/as-left calibration records.

Some stand-alone calibration management software is used only to schedule calibration and to store as-found/as-left calibration records.

Intelligent device management (IDM) software is another type of software that has a module for calibration, but it can work hand in hand with a CMMS for the work-order management part. IDM software can have a direct network connection to the field instruments to take advantage of built-in instrument diagnostics if they exist. Some instruments have a means to detect when calibration drift may have occurred, and they can alert and automatically trigger a work-order ticket in the

CMMS. Calibration might be called for earlier than originally scheduled or might be postponed until later.

ISA-TR108.1-2020, *Intelligent Device Management – Part 1: Concepts and Terminology* [7], provides concepts, definitions, and more information about IDM.

2.7 Calibration Equipment

The calibration program must consider how the test equipment is selected and managed. This section summarizes some of the considerations; these are discussed in more detail in Chapter 4.

2.7.1 Accuracy

Test equipment used to calibrate a device or a loop must be significantly more accurate than the device or loop that it is used to calibrate.

Refer to the calibration pyramid in Figure 2.3, and consider the three bottom layers. The field instruments are calibrated using the field calibrator (plant working standard). The field calibrator will be checked with a plant standard of higher accuracy, or it will be sent to a calibration laboratory to be certified. In this example, if the loop tolerance is ±1%, then the calibration equipment should have an accuracy of ±0.25%.

When assessing the accuracy of test equipment, consider the environment where the calibration equipment will be used. The temperature and other environmental factors can affect the accuracy of electronics in the equipment, just as they do with instrumentation. For example, if the equipment is used outdoors, the accuracy can be affected by high or low ambient temperature.

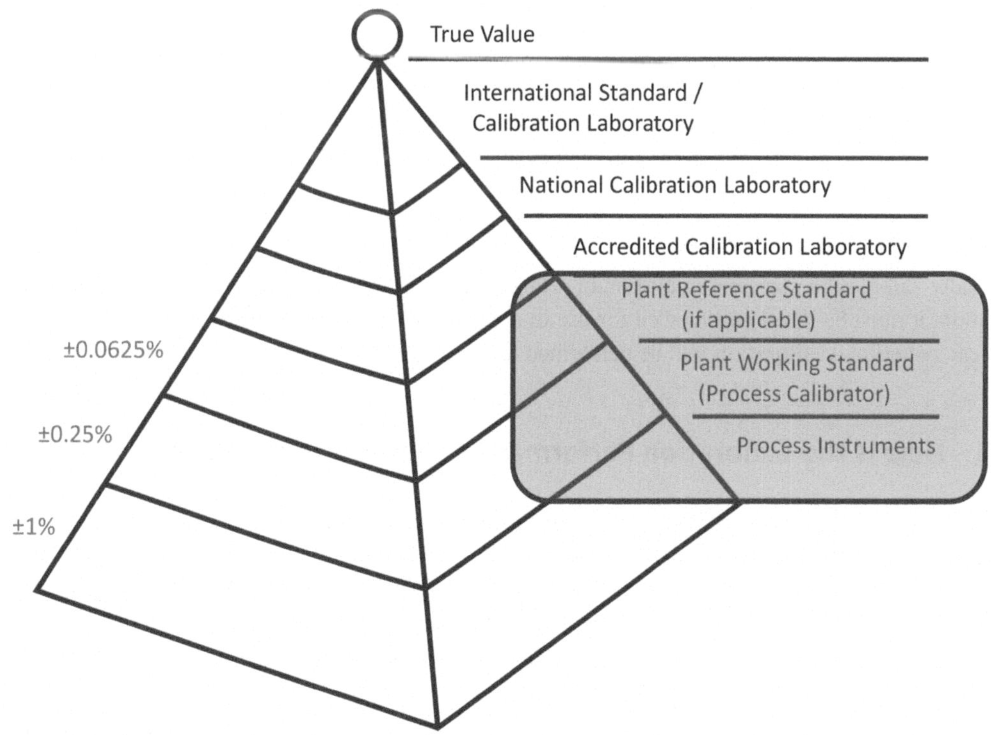

Figure 2.3 Test equipment example.

CALIBRATED			
DEVICE ID			
BY		DATE:	
NEXT CAL. DUE			

Figure 2.4 Calibration sticker.

If multiple devices are required to complete a calibration, then the total combined error of all calibration devices must meet the test uncertainty ratio (TUR) required. For example, if a calibration requires using a pressure reference source and a multimeter, combine the accuracies using the RSS method to calculate the TPE of the combination.

2.7.2 Certification

Test equipment should be certified regularly as recommended by the manufacturer. The equipment used to calibrate loops can be checked against the plant reference standard or sent to a calibration laboratory.

Once calibration equipment is certified, it should not be adjusted. There should be a means, such as a tamper-proof seal, to prevent or detect adjustments.

A sticker (Figure 2.4) should be affixed to the test equipment to document the most recent recertification date and when it is due for the next recertification. The information from the sticker will be recorded on the loop calibration record.

2.7.3 Environmental Conditions

Calibration equipment must be corrected for the effects of local conditions, such as atmospheric pressure, temperature, and local gravity. For example, the gravitational constant can vary at different locations on the planet, so a mass calibrator that is calibrated at a lab might need a correction factor for the local gravity. See the example in Section 6.5.2.

2.7.4 Safety

Consider safety when selecting equipment. For example, if a device will be used in a classified location, it must be listed and labeled for use in a hazardous area. Calibration procedures should indicate whether a calibration will be performed in a hazardous area.

2.8 How is the Calibration Performed? Calibration Procedures

In addition to inventorying instruments, determining calibration parameters and intervals, and purchasing appropriate test standards, a calibration program includes written procedures (work instructions) for performing calibrations. ISA-RP105.00.01-2017 [1] recommends developing a procedure and task list specifically for each loop or component to promote consistent results.

Calibration procedures should include the following:

- Description of the loop or component to be calibrated
- Overview of the task to be performed

- Equipment and personnel requirements
- Steps to be followed to perform the calibration
- Associated drawing references (e.g., loop drawings, location drawings, and data sheets)

If the calibration is to be done in situ, then the procedure should also include this information:

- Specifics regarding the process conditions
- Details about the interaction of the loop being calibrated/tested and the process
- Interactions such as alarms, interlocks, control points, or other aspects related to the loop
- Safety aspects such as energy isolation, personal protection equipment (PPE), and hazardous area classification

A few example procedures are included in Appendix D.

Calibration procedures can be written documents or electronic procedures on a computer or handheld device. The procedure and the record system might be incorporated into the same software application and database.

A manufacturer's technical manual is typically provided for each similar instrument model. Calibration instructions for the instrument are usually given in the technical manual. These instructions can be adopted or adapted as the calibration procedure for the applicable instruments. Most often, the manufacturer's calibration procedure is used to develop a company calibration procedure, which is approved by management and the quality department. However, if approved, the technical manual procedure may be used or referenced as the calibration procedure.

Generic calibration procedures can be developed for each instrument type. For example, one procedure could be developed for calibrating pressure instruments such as gauges and transmitters for pressure, differential pressure, and vacuum. Generic procedures should recommend using the manufacturer's technical manual to make necessary adjustments.

What information should be included in a calibration procedure? First, the format should follow the format required by your company's procedures. Also, any governing documents, such as a calibration policy, and any applicable procedures subordinate to a calibration policy must be followed.

The calibration procedure typically includes the following sections:

- **Purpose** – Clearly states the reason for the procedure, such as, "The purpose of this procedure is to provide standardized instructions for calibrating temperature instruments."
- **Scope** – Clearly states to what and to whom the procedure applies, such as, "This procedure applies to calibrating loop T-101 at the ABC Company, calibrated by employees and contract technicians."

Note: The purpose and scope can be combined into one section of the procedure. The information in these sections is typically obvious to those who perform calibrations, but managers need it to know what they're approving.

- **Definitions** – Contains brief descriptions of key terms, as applicable, for clarity. Acronyms and abbreviations used in the procedure are noted in this section to document their meaning throughout the text.
- **References/Attachments** – Identifies other documents, including attachments, that must be used with the procedure or that provide further information regarding the procedural content.
- **Test Equipment/Materials Required** – Identifies the test equipment and materials required to perform the procedure. Listing specific test equipment in this section helps to ensure uncertainty requirements are met and/or the desired accuracy ratio is achieved, particularly if the minimum tolerance achievable is specified in the scope or title of the procedure. Note that if the specified

test equipment is not available, the technician must notify the supervisor before performing the calibration.

- **Safety** – Provide information on potential human health hazards and potential hazards to the facility, equipment, or process. List safety work permit requirements. Include references to Safety Data Sheets (SDSs) for the material or chemicals involved.
- **Prerequisites/Initial Conditions (optional)** – Provides any conditions that should be met before performing the calibration, such as tank drained, controller in manual, or system shutdown. Alternatively, these conditions can be included in the test procedure and/or notes printed on the calibration data sheet.
- **Test Procedure** – This section outlines the procedure in a clear, concise, step-by-step manner. If any steps of the calibration procedure cannot be performed as specified, the technician must return the instrument to a safe condition and notify the supervisor.
- **Verification** – The pass/fail criteria may be included in the Test Procedure verification section or as a separate step at completion, evaluating the results obtained against the tolerances specified.
- **Approvals** – This section should include the signatures of the author and those authorized to approve the procedure.

2.9 Calibration Record System

Good calibration records are required to demonstrate product quality control, ensure safety, and comply with regulations and contracts. Most of those assurances are provided by accurate control and measurement loops. Device accuracy deteriorates during use. The only way to keep track of accuracy at any given time is to verify, adjust, and document the calibration data.

The calibration service record (CSR) is the form or database where the as-found and as-left calibration data is recorded when a calibration is performed. See the example form from ISA-RP105.00.01-2017 [1] in Appendix D.

Calibration data sheets may be hard copy, usually printed from a calibration software package. Some are totally manual with no preprinted information. There are also paperless calibrations where the data is either automatically collected by a documenting calibrator or manually entered into a handheld device as an electronic record. In any case, the following information should be included as part of a CSR form. Figures 2.5 through 2.13 are clipped from the form in Appendix D to show how to complete the standard ISA form.

- **Loop or device tag number** (Figure 2.5) – This is a unique identifier or unique combination used as the main tracking number for each instrument. It will come from the instrument index or P&ID.
- **CMMS identifier** –If a CMMS is used, the device tag will be associated with a unique number in that system. (*Note*: This is not shown on the ISA example form.)
- **Device identification** – This is an additional identification (ID) number that is sequentially assigned to a device as an additional tracking number. The tag number and CMMS identifier

Instrument/Loop Tag: **PIT-303 / PIC-303**
IO Point ID: **CTB1 A8 - PIC-303**

Figure 2.5 The tag number of the instrument or loop and the input-output point identification. © International Society of Automation. Used with permission of ISA.

reference the instrument location within a system, but the device identification number stays with the device, like a manufacturer's serial number. This way, the history of any instrument installed in the instrument location (tag number) is traceable, and the history of any instrument is traceable. Many instruments stay in the same location for the life of the equipment, in which case, this is not as important. However, failures occur, and instruments must be replaced. In other instances, spares are installed temporarily to keep a process running when an instrument is removed for bench test calibration.

This traceability is important, for example, in a pharmaceutical facility that produces penicillin. Any instrument removed from a penicillin-manufacturing area cannot be used in any other manufacturing area. The instrument identification number used to track where an instrument is or has been is useful in ensuring that the instrument is not installed in the wrong location. It is a useful practice to use both types of identification numbers. (*Note*: This is not shown on the ISA example form.)

- **Nameplate data** (Figure 2.6) – The manufacturer, model number, and serial number should be listed on the calibration data sheet.
- **Calibration range and calibration tolerance** (Figure 2.7) – This defines the upper and lower limits used for calibration. Ideally, this is the input and output range, if applicable. For example, a good format for the calibration range of a temperature transmitter would be 0 °C to 100 °C = 4 mA to 20 mA with a calibration tolerance of ±1.0 °C/±0.16 mA. The calibration range is not always the same as the instrument range or capability.
- **Location** (Figure 2.8) – Be as specific as possible about the location of the instrument so a technician doesn't waste time searching.
- **Calibration procedure number** (Figure 2.9) – This is the calibration procedure used to perform the calibration. In some facilities, the entire calibration procedure is printed on the calibration data sheet.

Plant:
Operating Unit:
Location:
Manufacturer: **Rosemount**
Model: **2088 G 3 S**
Serial #: **(record serial number)**

Figure 2.6 The manufacturer, model, and serial number fields. © International Society of Automation. Used with permission of ISA.

	Input	Control System Indication
LRV	**0**	**0**
URV	**200**	**200**
Eng Units	**psig**	**psig**

Loop/device required tolerance **±2 psig**

Figure 2.7 The calibration range and tolerance levels. © International Society of Automation. Used with permission of ISA.

Calibration Location (in-situ, shop, etc.): **In situ, loop calibration**

Figure 2.8 The calibration location. © International Society of Automation. Used with permission of ISA.

Special Instructions: **Calibration procedure SOP-CAL-01**

Figure 2.9 The calibration procedure number location on the calibration record. © International Society of Automation. Used with permission of ISA.

Calibration Date: **Nov 1, 2024**

Next Calibration Due: **Nov 1, 2025**

Figure 2.10 The calibration dates. © International Society of Automation. Used with permission of ISA.

Input %	Applied Input (Standard)	Desired Indication	As Found						As Left					
			Increasing Indication	Error in EU	Error in %	Decreasing Indication	Error in EU	Error in %	Increasing Indication	Error in EU	Error in %	Decreasing Indication	Error in EU	Error in %
0														
25														
50														
75														
100														

Figure 2.11 The as-found and as-left data sections on the form. © International Society of Automation. Used with permission of ISA.

- **Last calibration date, calibration due date, and calibration interval** (Figure 2.10) – This information should be included to ensure the calibration is performed periodically as required.
- **As-found data and as-left data** (Figure 2.11) – This data relates the specified test points to the corresponding test standard value. Calibration programs serve the dual purpose of checking loops for conformance to tolerance requirements and, if tolerance is not met, making adjustments to equipment to bring it into tolerance. If all as-found data is within tolerance and no adjustments are made, the as-left data would be N/A or the same as the as-found data. Every effort should be made to record as-found data for failed instruments before making any adjustments to provide the most data for evaluation. When documenting calibration information (i.e., as-found and as-left), calculate and document the accuracy of the observed indications. Ensure that the calculated accuracies are expressed in the same manner as the required loop/device tolerance (e.g., engineering units, percentage of full scale, percentage of reading, or engineering units) to enable direct comparison.
- **Test standards** (Figure 2.12) – Record the unique identification of any test standards used to perform the calibration, and record the calibration due date of the standard(s). It would be best to record this before beginning the calibration to ensure each standard is within its calibration periodicity (but, of course, you checked this when you obtained the standard from the shop).

Test Equipment						
Manufacturer	Model	Serial	Certification No.	Certification Date	Accuracy (>4x loop / device)	Remarks
Fluke	729	xxx	xxx	1-Oct-24	±0.02% FS = 0.06 psig	use 300 psi module

Figure 2.12 The test equipment section. © International Society of Automation. Used with permission of ISA.

Special Instructions: **Calibration procedure SOP-CAL-01**	
Comments:	
Calibrated by:	Date:

Figure 2.13 The comments and sign-off sections. © International Society of Automation. Used with permission of ISA.

ABC Company – Calibration Status
ID: ____________
Cal Date: ____________
Cal Due: ____________
Technician: ____________

Figure 2.14 Calibration status label example.

The most important reason for documenting the test standards is for reverse traceability in case a standard is found to be out of tolerance. If a test standard is found out of tolerance on its next calibration, then any calibration performed using that standard since its last known calibration should be evaluated to determine a course of action.

- **Comments, technician signature, date of calibration** (Figure 2.13) – Record any comments or observations, date the form, and sign off.

In addition to the calibration record system (CRS) form, calibration status labels are used to provide a visual indication of the calibration status of an instrument. Various label styles are used across different industries. The following information must be displayed to show the calibration status: the instrument identification (such as the tag number, instrument ID number, or serial number), calibration date, next calibration due date, and the technician who performed the calibration (e.g., initials and employee ID). An example is illustrated in Figure 2.14.

2.10 Who Performs Calibrations – Staff Qualifications and Training

2.10.1 The Control System Technician

A control system technician (CST) is a skilled craftsperson who knows pneumatic, mechanical, and electrical instrumentation. A CST understands process control loops and process control systems, including those that are computer-based. Typically, a CST has received training in such specialized subjects as theory of control, analog and digital electronics, microprocessors, computers, and the operation and maintenance of field instrumentation. ISA's CCST program lists the tasks that a CST should be able to perform.

A CST performs calibration, documentation, loop checks, troubleshooting, and instrumentation adjustment, repair, or replacement. These tasks relate to systems that measure and control level, temperature, pressure, flow, force, power, position, motion, speed, mass, physical properties, chemical composition, and other process variables.

Only competent personnel should be allowed to perform calibrations. A CST's qualifications can be confirmed through an internal company certificate program or a recognized independent, third-party certification program. A subject matter expert should evaluate a CST's abilities and qualifications. Requirements for periodic training and recertification should be considered. Regulations in particular industries might drive some of the training and certification requirements.

2.10.2 Characteristics of a CST

The following are characteristics of a good CST:

- Honesty and integrity
- Attention to detail
- Excellent documentation practices
- Understanding of processes

2.10.2.1 Honesty and Integrity

A CST must possess honesty and integrity above all else. Most technicians work independently much of the time. The calibration department may be understaffed, and production schedules may demand unrealistic completion requirements. These factors can have a real impact on proper performance and documentation of calibrations. Remember: Nobody can take away your integrity; however, you *can* give it away. Calibrations must be performed according to the procedures and properly documented.

2.10.2.2 Attention to Detail

Calibrations should be performed according to the detailed instructions. Each make and model of an instrument is adjusted differently. Each instrument is installed in a different physical and loop configuration. Because of these and many other differences, attention to detail is important. The minute a technician is not paying attention to detail, safety and proper performance are jeopardized.

2.10.2.3 Excellent Documentation Practices

In many facilities, the impression of quality is determined by the content and appearance of documentation; documentation is essential to meet quality standards. For example, pharmaceutical companies must follow the Food and Drug Administration's Current Good Manufacturing Practice regulations. If it isn't documented, it wasn't done.

Calibration data sheets must be neat, complete, signed, and, if required, reviewed in a timely manner. Also, all related documentation, such as drawings, manuals, specifications, and databases, must be updated when changes occur.

2.10.2.4 Understanding Processes

One thing that sets technicians apart is their understanding of the process, particularly how the instruments monitor and control the process. There is a difference between calibrating an individual component and calibrating an instrument as part of the bigger process control loop. For example, knowing when a controller can be placed in manual without affecting the process and what to do while that controller is in manual requires an understanding of the process.

Additionally, when an operator says there is a problem with the indication, a technician who knows the instrument loop and process will be more capable of identifying the cause of the problem.

2.11 Loop and Component Performance Verification

A CST will be responsible for carrying out the calibration procedures at the intervals defined in the calibration program. It is important to follow all the procedure steps shown in the flowchart in Figure 2.15.

When a loop accuracy is not within the required tolerance, the procedure should indicate whether the device is to be adjusted or replaced. Some troubleshooting might be necessary. If the required accuracy cannot be achieved with adjustment, the device should be replaced.

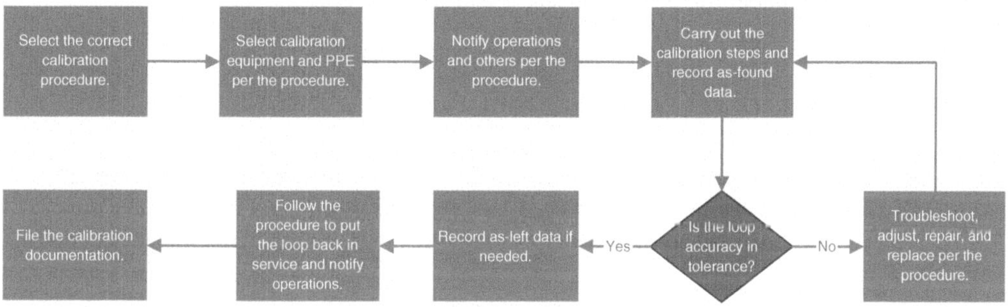

Figure 2.15 Performance verification flowchart. PPE, personal protective equipment.

2.12 Safety Considerations

It is important to perform calibrations safely. Product safety is also of concern when performing calibrations in a manufacturing environment. Many resources on safety are available from ISA, including *Troubleshooting: A Technician's Guide* [8].

Here are a few things calibration technicians can do to improve safety in day-to-day work activities.

- Include specific safety considerations in each calibration procedure.
- Inform operations when performing calibrations in situ.
- Keep the shop and work areas clean and free of trip hazards.
- Work with a partner, or at least make sure someone knows where you are working at all times.
- Some instruments are always installed in difficult places to reach. If it is possible to install a permanent platform, then have it done. Otherwise, use safety harnesses, ladders, and lifts properly.
- If an instrument is installed in a classified area, then the test equipment must be rated for the area. If the procedure requires opening covers of explosion-proof devices, then a hot work permit might be required.
- Technicians may be exposed to lethal electrical voltages. Know the high voltage areas, de-energize electrical circuits that are not required, and use proper electrical safety practices

(e.g., an insulated floor mat, rubber electrical safety gloves, a roped-off area, or a safety man outside the area with a rope tied around the CST).
- Ensure electrical power cords are properly insulated. Ensure equipment is properly grounded.
- If you're not sure, then ask!

2.13 Summary

This chapter presented the key aspects of a calibration program consistent with ISA-RP105. Many of the CCST Level III tasks relate to planning and administering the calibration program. It is also important for the Level I and Level II CCSTs and all other CSTs involved in calibration activities to understand these concepts and how a calibration program is organized.

Key concepts that were covered include

- calibration schedule and tolerance based on criticality,
- selecting and maintaining calibration equipment, and
- performing and documenting calibrations safely and with integrity.

Review Questions

2.1 Which of the following is the primary reason for implementing a calibration program?
a) To ensure all instruments are replaced on a regular schedule
b) To detect and correct instrument errors to ensure measurement accuracy to the required tolerance
c) To reduce the need for preventive maintenance
d) To allow operators to make manual adjustments based on plant conditions

2.2 What is the first step in establishing a calibration program?
a) Set the calibration frequency for each loop.
b) Set the required tolerance for each loop.
c) Establish a tagging system and create a list of loops or devices that require calibration.
d) Write calibration procedures.

2.3 If loops associated with regulated emissions monitoring must be calibrated on a specific schedule, then what is a recommended method of classifying them?
Think of some examples in your work where criticality classification could be applied.

2.4 Why is it important to assign a criticality classification to instruments in a calibration program?
a) To determine which instruments require protective enclosures
b) To prioritize calibration frequency, capture regulatory requirements, or identify maintenance requirements for each loop
c) To ensure that all instruments are calibrated at the same interval, regardless of function or importance
d) To classify instruments by manufacturer for easier spare parts management

2.5 A custody transfer contract specifies that the product flow measurement must be accurate to ±2% of the meter range. This information would be used to specify what about the loop calibration?
a) Loop tolerance
b) Meter span, upper range value (URV), lower range value (LRV)
c) Meter accuracy
d) Loop confirmation interval

2.6 If the meter instrument range in the previous question is 0–1,000 gal./hr., what is the loop tolerance in units of measure?

2.7 What is the purpose of establishing a calibration confirmation interval for a loop?
a) To match the instrument's warranty period
b) To determine how frequently a sensor must be replaced
c) To verify that the instrument continues to perform within tolerance between calibrations
d) To allow operators to bypass alarms during normal operation

2.8 Calibration confirmation intervals are established for each loop to define how often the loop must be calibrated. How would a reliability-centered maintenance program modify the confirmation intervals?

2.9 Which document provides step-by-step instructions, acceptance criteria, and equipment required to calibrate a loop?

2.10 Which document is used to record the results of a calibration, such as the as-found and as-left readings?

References

Note: The dates and editions for standards and other references in this chapter are current as of publication. The reader is encouraged to check for later revisions or editions.

1 International Society of Automation (ISA), ISA-RP105.00.01-2017, *Management of a Calibration Program for Industrial Automation and Control Systems.* Research Triangle Park, NC: ISA, 2017.
2 American National Standards Institute/International Society of Automation (ANSI/ISA), ANSI/ISA 5.1-2009, *Instrumentation Symbols and Identification.* Research Triangle Park, NC: ISA, 2009.
3 Instrument Society of Automation (ISA), ISA-5.4-1991, *Instrument Loop Diagrams.* Research Triangle Park, NC, ISA, 1991.
4 Meier, Frederick, and Clifford Meier, *Instrumentation and Control Systems Documentation*, 2nd ed. Research Triangle Park, NC: Instrumentation, Systems, and Automation Society (ISA), 2011.
5 Instrumentation, Systems, and Automation Society (ISA), ISA-TR91.00.02-2003, *Criticality Classification Guideline for Instrumentation.* Research Triangle Park, NC: ISA, 2003.

6 Emerson, 2021, March, "How to Calculate Pressure Transmitter Calibration Intervals," Technical Note 00840-1900-4801, Rev AA. Emerson. https://www.emerson.com/documents/automation/technical-note-how-to-calculate-pressure-transmitter-calibration-intervals-en-7432098.pdf. Accessed February 3, 2026.

7 International Society of Automation (ISA), ISA-TR108.1-2020, *Intelligent Device Management – Part 1: Concepts and Terminology*. Durham, NC: ISA, 2020.

8 Mostia, William L., *Troubleshooting: A Technician's Guide*, 2nd ed. Research Triangle Park, NC: International Society of Automation (ISA), 2006.

3

Control System Documentation

3.1 Piping and Instrumentation Diagrams

Piping and instrumentation diagrams (P&IDs) provide a detailed overview of a process system (see the example in Appendix A). They include major components, utilities, flow paths, supporting equipment, and instrumentation.

Although P&IDs are commonly used in process industries, there is no standard P&ID. Most companies have their own P&ID standards that are based on industry guidelines. It is worthwhile for a control system technician (CST) to be aware of these.

Process Industry Practices (PIP; www.pip.org) has developed a P&ID practice, PIC001 *Piping and Instrumentation Diagram Documentation Criteria* [1]. ISA standard ANSI/ISA-5.1-2024, *Instrumentation and Control – Symbols and Identification* [2], is the most generally accepted guide for symbolism on instrumentation and control systems.

From the P&ID of a particular system, a technician can determine pertinent information about the instrumentation and controls applicable to performing calibrations and understanding the system operation, including:

- Components of an instrument loop.
- Functional identification.
- Methods of signal transmission (e.g., pneumatic, electronic, hydraulic, and software link).
- Controller input(s) and output(s).
- Control valve characteristics (fail position and direct/reverse-acting).
- Flow sensor types.
- Process isolation valves.
- Process fluid.

P&IDs are also used to organize project documentation. Using P&IDs as the base for all information in a large project provides a single reference point for data provided in other documents. This makes sense because P&IDs are used to define system boundaries. P&IDs must be controlled and maintained up to date. During start-ups, the master P&IDs are marked up and highlighted to reflect the state of a system at any time. If a technician discovers any discrepancy, they must take responsibility to ensure the controlled drawing is properly revised.

An example of a P&ID is in Appendix A. Look at the P&ID and see what can be determined about the temperature loop (Figure 3.1) for this process.

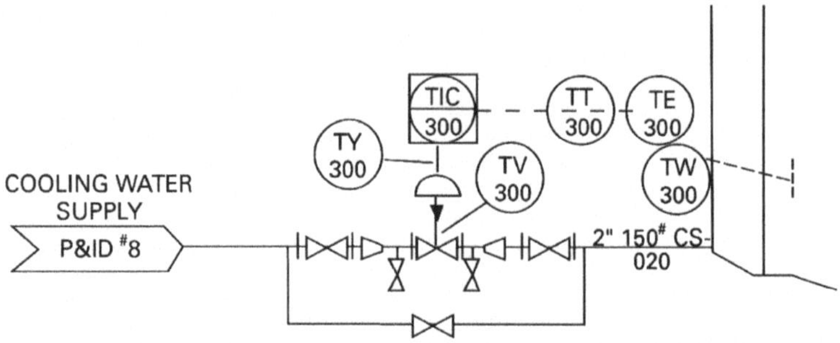

Figure 3.1 T-300 loop from the P&ID in Appendix A. TE, temperature element; TIC, temperature indicating controller; TT, temperature transmitter; TV, temperature valve; TW, thermowell; TY, temperature auxiliary device.

The P&ID (Figure 3.1) shows that the temperature loop includes TW-300, TE-300, TT-300, TIC-300, TY-300, and TV-300. Using ANSI/ISA-5.1-2024 [2] as a reference:

- TW in the loop is a temperature well, or thermowell, that provides a physical boundary between the temperature sensor and the process. The temperature sensor, TE, is inserted into the TW. The TW is not a component in the signal conversion of the temperature signal or control. It is important for the CST to know that there is a TW in place so the TE can be removed for calibration.
- TE is a temperature sensor, but it is not indicated whether the sensor is a resistance temperature detector (RTD) or a thermocouple. That would be determined from the data sheet or physical inspection.
- TT is a temperature transmitter that provides the input to TIC-300.
- TIC is a shared display, a shared control temperature indicating controller.
- TY is between the electronic output of the TIC and the pneumatic input of the valve, TV. According to Table 1 of ANSI/ISA-5.1-2024 [2], the *Y* in TY refers to relay, compute, or convert. In this case, the TY is a conversion device in the temperature loop. Most likely, the device is an I/P transducer that converts a 4–20 mA signal from the TIC to a 3–15 psig signal for the valve.

3.2 Loop Diagrams

A loop is a combination of interconnected instruments that measures and/or controls a process variable. An instrument loop diagram is a composite representation of instrument loop information containing all associated electrical and piping connections. Instrument loop diagrams are developed according to ISA-5.4-1991, *Instrument Loop Diagrams* [3]. Example loop diagrams are included in the reference section, Appendix B. The minimum content requirements include:

- Identification of the loop and loop components, including sensors, transmitters, controller inputs, control logic, controller outputs, conversion devices, and final control elements.

- Point-to-point interconnections with identifying numbers or colors of electrical and/or pneumatic wires and tubing, including junction boxes, terminals, bulkheads, ports, and grounding connections.
- General location of devices, such as in the field, panel, input/output (I/O) cabinet, and control room.
- Energy sources of devices.
- Control action or fail-safe conditions.

Additional content requirements, format, and examples are included in ISA-5.4-1991 [3].

The information on the loop diagram is important for writing calibration procedures that involve putting loops in manual control or taking them out of service during calibration. They are also an aid in troubleshooting when a loop tests out of tolerance.

A formalized loop check should be documented before placing any loop in service. This formalized program should include verification of installation against the loop diagram and simulation of signals to verify output responses and indications throughout the range. Why is this important? A significant percentage of instrument loops have problems that may result in hidden failures. For example, a temperature transmitter output wired to a programmable logic controller (PLC) analog input provides a temperature display on an operator interface. The transmitter is calibrated from 0 °C to 100 °C to provide a proportional 4–20 mA output. If the PLC programmer writes the code for this input as 0 °C to 150 °C for a 4–20 mA input and a loop check is not performed, an inaccurate displayed value will go undetected. Other typical problems found and corrected by performing loop checks include wiring connected to the wrong points, ground loops, and broken wires. Loop checking is another certified control system technician (CCST) domain. For more information, refer to *Loop Checking: A Technician's Guide* [4].

3.3 Instrument Specification Forms/Instrument Database

Instrument specification forms contain the information necessary to obtain vendor quotes and purchase instrument devices. ISA-TR20.00.01-2007, *Specification Forms for Process Measurement and Control Instruments* [5], includes device specification forms for many instruments that can be completed for this purpose. The information contained on each form is specific to the instrument type but typically includes dimensions, materials of construction, design temperature/pressure, connection sizes, ambient conditions, indicator details, instrument ranges/tolerances, and so on.

When instruments are received from the vendor, each should be verified against the specification to the maximum extent practicable. This should be documented using a formalized receipt verification process.

For facilities that do not have instrument specification forms from the system design, these forms can be developed using ISA-TR20.00.01-2007 or according to the company procedure [5]. Some calibration departments develop their own forms to specify the information needed to perform calibrations. Users and the quality department should review and approve these specifications to ensure process and quality requirements are considered.

Instrument details might be kept in an instrument database. Instrument database software usually has a feature to produce a specification form for an instrument.

3.4 Project Specifications

Any new manufacturing facility or large expansion project will have a set of project specifications. They include architectural, plumbing, mechanical, electrical, and instrumentation specifications. For instrumentation, the project specifications are usually general but include useful content such as:

- Instrument tagging requirements.
- Typical installation detail drawings by instrument type.
- Size and type of pneumatic tubing.
- Approved methods of mounting instruments.
- Wire labeling conventions.
- Intrinsic safety standards.

For some facilities, the project specifications may include detailed instrument specification forms and an original master instrument list or database.

3.5 Manufacturer's Specifications

The manufacturer's technical manual or product literature almost always has a specification section. Understanding what the manufacturer's specifications mean is important when purchasing an instrument that is acceptable for the process requirements and to the technician maintaining the instrument.

Examples of manufacturer's instrument specifications are included in Appendix C. This instrument data is needed when creating calibration procedures and can be a useful reference when performing calibrations. This information can also be used when completing an ISA-style form, such as those in ISA-TR20.00.01-2007.

The most important specification for the calibration technician is the instrument range and accuracy. The range capability of the instrument is usually understandable; accuracy is another matter. For example, accuracy may be specified as 0.25%. What does this mean? Is it 0.25% of the range or the reading? For calibrating a 0 to 100 psig pressure transmitter at 20 psig, this can mean the difference between 0.25 psig and 0.05 psig.

In addition to the specified base accuracy, the manufacturer will provide other information, such as long-term stability, effects of ambient conditions, effects of process conditions, and installation effects on overall performance. The specification might give repeatability with a confidence interval instead of accuracy. If it is not clear, ask a technical resource from the manufacturer what the specification means.

3.6 Summary

This chapter has described the most common documentation utilized by the calibration technician. Table 3.1 is a summary of the information available in each document.

Table 3.1 Sources of calibration information.

Document	Calibration Information
P&ID	Components of an instrument loop Functional identification Methods of signal transmission Controller input(s) and output(s) Control valve characteristics (e.g., fail position and direct/reverse-acting) Flow sensor types
Loop diagram	Identification of the loop and loop components Point-to-point interconnections General location of devices Energy sources of devices Control action or fail-safe conditions
Instrument specification sheet	Specified calibration ranges and units of measure Specified output ranges
Project specifications	Instrument tagging requirements Typical installation detail drawings by instrument type Size and type of pneumatic tubing Approved methods of mounting instruments Wire labeling conventions Intrinsic safety standards
Manufacturer's specifications	Instrument range Accuracy Stability
Calibration procedures	Safety information Step-by-step procedures Calibration equipment requirements
Calibration record sheet	As-found data As-left data Calibration equipment information

Review Questions

Matching exercise: Choose the correct document(s) associated with each item of information in this list. There might be more than one answer for each item. Some of these are covered in Chapter 2.

Which output device does the instrument control?
Is the temperature element installed in a thermowell?
What steps should be followed when calibrating a loop?
Where should the results be recorded?
What is the configured range of a device (lower range value [LRV], upper range value [URV])?
What is the maximum range of the device (lower range limit [LRL], upper range limit [URL])?
What are the base accuracy, stability, and other effects on uncertainty?
What is the allowable loop tolerance?

Documentation:

1) Piping and instrumentation diagram (P&ID)
2) Loop diagram
3) Instrument specification sheet
4) Manufacturer's data sheets
5) Calibration procedure
6) Calibration service record

References

Note: The dates and editions for standards and other references in this chapter are current as of publication. The reader is encouraged to check for later revisions or editions.

1 Process Industry Practices (PIP), PIC001, *Piping and Instrumentation Diagram Documentation Criteria*, 2026 Technical Revision. Austin, TX: PIP.

2 International Society of Automation (ISA), ANSI/ISA-5.1-2024, *Instrumentation and Control — Symbols and Identification*. Durham, NC: ISA, 2024.

3 Instrument Society of America (ISA), ISA-5.4-1991, *Instrument Loop Diagrams*. Research Triangle Park, NC: ISA, 1991.

4 Jeffrey, Harley M., *Loop Checking: A Technician's Guide*. Research Triangle Park, NC: The Instrumentation, Systems, and Automation Society, 2005.

5 Instrumentation, Systems, and Automation Society (ISA), ISA-TR20.00.01-2007, *Specification Forms for Process Measurement and Control Instruments – Part 1: General Considerations*. Research Triangle Park, NC: ISA, 2007.

4

Calibration Equipment and Systems

4.1 What Does Field Calibration Equipment Do?

Performing a complete loop calibration requires using equipment that is designed to be used outside the laboratory. The equipment will have the following functions:

- Apply simulated process conditions to a sensor (e.g., apply a specified temperature to a resistance temperature detector [RTD] or thermocouple)
- Measure a sensor output (e.g., RTD resistance, thermocouple voltage, or a transmitter 4–20 mA signal)
- Simulate a sensor output or transmitter input (e.g., millivolts to simulate a thermocouple)
- Compare a standard reference to the equipment being tested (e.g., use a standard reference device to measure the applied pressure and compare it to the pressure indicated by testing the loop)

4.2 Types of Calibration Standards

Calibration equipment can be designed for a laboratory environment, or it can be portable to carry to the device for in situ calibration. Laboratory equipment usually has more functionality and higher accuracy than portable equipment, but it might have limited operating environment specifications. Portable equipment usually runs on batteries and is small enough to carry out to the field and calibrate a device at its location.

Documenting calibrators have more software options and features geared for calibration needs. Some of these can run automated test sequences, record results, control other test and calibration instrumentation, and allow for entering notes. They can connect to data systems to upload calibration results and to download procedures and data sheets.

4.3 Selecting Equipment

For each calibration, the certified control system technician (CCST) must ensure that the test equipment is adequate for the calibration task. According to ISA-RP105.00.01-2017: "Equipment used to evaluate a loop's performance against its required tolerance or to calibrate a device should be certified, accurate, corrected for local conditions and have adjustments controlled" ([1], Clause 5.7).

- The calibration reference standards should be traceable to national or international standards (see Section 1.4). Equipment used in field calibrations can be certified by a qualified calibration laboratory. Equipment should be certified periodically and have a sticker attached that documents the most recent certification date. The sticker should also indicate the next due date of certification. Equipment that is overdue for certification should not be used for calibration.
- The test equipment accuracy should be related to the required tolerance of the loop or device being tested. It should have a test uncertainty ratio (TUR) ≥ 4 (see Section 4.4). If multiple calibration equipment components are used to perform the calibration, the uncertainties should be combined using the root sum squared (RSS) method.
- Equipment should be identified with a unique identification number, part number, and range/accuracy. This information will be recorded on the calibration record sheet along with the certification details.
- Equipment should be corrected for the effects of local conditions such as temperature, atmospheric pressure, and local gravity.
- Equipment should have a means to control access to adjustments affecting performance. These should be sealed to prevent unauthorized changes. No adjustments should be made in the field to equipment that has been calibrated and certified.

4.4 Is the Calibration Equipment Accurate Enough?

CCST Level III

4.4.1 Calculating TUR

ISA-RP105.00.01-2017 recommends using the TUR to determine whether a reference standard is accurate enough for the test [1]. TUR is the ratio of the tolerance of the device (or loop) under test to the total uncertainty of the measurement system used to calibrate it (Equation 4.1).

$$\text{TUR} = \frac{\text{tolerance of device or loop}}{\text{total uncertainty of the calibration equipment}} \tag{4.1}$$

The recommended practice is to have a TUR ≥ 4. The calculation for TUR is shown in Equation 4.2.

$$\text{TUR} = \frac{\text{USL} - \text{LSL}}{2u} \tag{4.2}$$

USL = upper specification limit of the loop/device tolerance
LSL = lower specification limit of the loop/device tolerance
u = uncertainty of the measurement process used for calibration

The USL and LSL are calculated from the tolerance specified for the loop or device calculated at the measurement value. The value for u is the total probable error (TPE) for all the calibration devices combined using RSS. Figure 4.1 illustrates the relationship between the calibration system uncertainty and the loop tolerance.

For the value of uncertainty (u), it is important to consider not only the reference standard device accuracy but also all sources of uncertainty in the calibration system. Use the TPE for the reference standard calculated at the test conditions to account for process and environmental conditions that affect uncertainty. TPE for the reference standard is calculated just as it is for a device under test

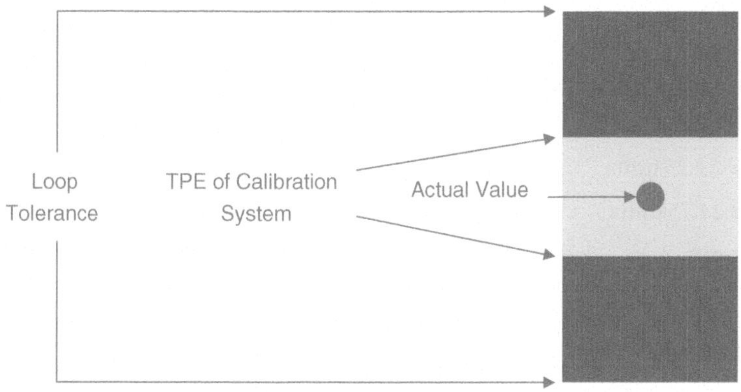

Figure 4.1 Illustration of TUR input values. TPE, total probable error.

(see Section 1.2.8 for the TPE calculation). If multiple devices are used for the calibration (such as a pressure measurement device and an ammeter), then the TPE calculation should combine the uncertainties of all of the calibration devices.

The result can differ depending on whether the tolerance is stated in terms of reading, span, or full scale.

Example 4.1 ***Percentage of full scale***

Loop tolerance = ±1% of full scale; full scale = 100 units.
TPE for the reference standard is 0.1% of full scale of 100 units.

$$\text{USL} = 100 + 100 \times 1\% = 101 \text{ units}$$

$$\text{LSL} = 100 - 100 \times 1\% = 99 \text{ units}$$

$$u = 100 \times 0.1\% = 0.1 \text{ units}$$

$$\text{TUR} = \frac{101 - 99}{2 \times 0.1} = \frac{2}{0.2} = 10$$

These values are illustrated in Figure 4.2.

This TUR would be acceptable because it is more than 4. It would apply to all test measurements because the tolerance is the same number of units over the full range of the test.

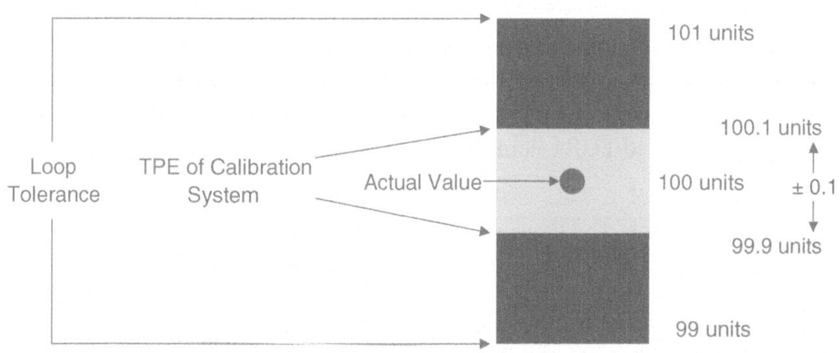

Figure 4.2 Example TUR illustration. TPE, total probable error; TUR, test uncertainty ratio.

Example 4.2 ***Percentage of reading***
Loop tolerance = ±1% of reading.
TPE for the reference standard is 0.1% of full scale of 100 units.

For a test point at 25% of the range:

$$USL = 25 + 25 \times 1\% = 25.25 \text{ units}$$

$$LSL = 25 - 25 \times 1\% = 24.75 \text{ units}$$

$$u = 100 \times 0.1\% = 0.1 \text{ units}$$

$$TUR = \frac{25.25 - 24.75}{2 \times 0.1} = \frac{0.5}{0.2} = 2.5$$

For a test point at 75% of the range:

$$USL = 75 + 75 \times 1\% = 75.75 \text{ units}$$

$$LSL = 75 - 75 \times 1\% = 74.25 \text{ units}$$

$$u = 100 \times 0.1\% = 0.1 \text{ units}$$

$$TUR = \frac{25.25 - 24.75}{2 \times 0.1} = \frac{1.5}{0.2} = 7.5$$

The TUR is acceptable for the 75% test point ($7.5 > 4$) but not the 25% test point ($2.5 < 4$).

4.4.2 Calculating Test Accuracy Ratio

Older standards and texts (now withdrawn) defined the term *test accuracy ratio* (TAR). It is mentioned here because some organizations still use this method to assess the reference standard. ISA-RP105.00.01-2017 [1] and other more recent standards recommend using TUR instead of TAR because TUR considers all sources of uncertainty in the calibration system, whereas TAR only considers the stated accuracy (Equation 4.3). TAR is simpler, but it can underestimate error in calibration because it does not consider uncertainty contributions like environmental effects on the calibrator. TAR is not formally defined in any current standards.

$$TAR = \frac{\text{tolerance of device/loop}}{\text{accuracy specification of measurement standard}} \tag{4.3}$$

Example:
Loop tolerance = ±1.0 unit.
Stated accuracy of the reference standard = ±0.1 units:

$$TAR = \frac{1.0}{0.1} = 10$$

This TAR would be considered acceptable because it is more than 4.

TAR is used in some of the examples in this text because some organizations still use it. It can also be used to do a quick screening to see if a reference standard comes close to being acceptable before bothering with the more detailed TUR calculation.

4.5 Adjusting for Local Conditions

Calibration equipment must be corrected for the effects of local conditions such as atmospheric pressure, temperature, and local gravity. For example, the gravitational constant can vary at different locations on the planet, so a mass calibrator that is calibrated at a lab might need a correction factor for the local gravity. See the example in Section 6.5.2 for a pressure calibrator. If the equipment isn't adjusted, the control system technician (CST) must account for the error when calculating the calibration equipment TPE. The calibration uncertainty is not the same as the device accuracy.

Smart equipment can make the adjustments automatically or with user input (e.g., tell the dead weight tester what the local gravity is).

4.6 Calibration Management Software Systems

Software with the features listed is available to help manage the calibration program.

- Store calibration procedures
- Record calibration equipment data and certification status
- Access instrument data sheets
- Perform TPE and TUR calculations
- Track calibration results
- Interface to smart calibration equipment to upload/download procedures, notes, and calibration results

This software is used for the calibration record system (CRS) described in Section 2.9.

Review Questions

4.1 What is one big difference between bench calibration equipment and field calibration equipment?

a) Field equipment is typically battery powered and portable.
b) Bench equipment must be certified more often.
c) Field calibration equipment is typically more accurate than bench equipment.
d) Bench equipment is used for mechanical instruments; field calibration equipment is used for electrical instruments.

4.2 What does the TUR represent in a calibration?

a) The ratio of the calibration equipment's TPE to its base accuracy
b) The ratio of the loop tolerance to the TPE of the calibration equipment
c) The time until the calibration equipment must be recertified
d) The ratio of the standard's accuracy to the ambient temperature

4.3 What is the recommended minimum TUR for most circumstances?

4.4 Calculate the TUR for this calibration at the full scale of 200 psig:

- Loop tolerance = ±1 psig
- Calibrator TPE = ± 0.5 psig

Is this TUR acceptable?

4.5 What is the purpose of having calibration equipment calibrated by a certified lab?

a) To ensure the equipment is new and unused
b) To reduce the need for documenting measurements

c) To ensure the measurement is traceable to national and international standards through an unbroken chain of calibrations
d) To enable faster measurements

4.6 How can a CST know the calibration equipment is within its certification interval?

4.7 Why is it important to know if a device must be adjusted for local conditions?
a) To prevent electronic interference
b) So the equipment can be used without a calibration procedure
c) Because temperature, atmospheric pressure, and local gravity can affect measurements

Reference

Note: The dates and editions for standards and other references in this chapter are current as of publication. The reader is encouraged to check for later revisions or editions.

1 International Society of Automation (ISA), ISA-RP105.00.01-2017, *Management of a Calibration Program for Industrial Automation and Control Systems.* Research Triangle Park, NC: ISA, 2017.

5

Temperature Instruments

5.1 What Is Temperature?

Temperature is a property of an object that determines the direction of heat flow when the object is in thermal contact with another object. Heat is thermal energy. Heat flows from a higher temperature to a lower temperature.

5.2 Temperature Terms and Units of Measure

Frequently used units of measurement for temperature are listed in Table 5.1.

Absolute zero temperature is the temperature at which all particle motion stops. It is a theoretical value that cannot actually be reached, but is a useful baseline.

Figure 5.1 shows the relationship of important temperature terms. The units Kelvin (K) and Rankin (R) are both units relative to absolute zero temperature. They will never have a value below

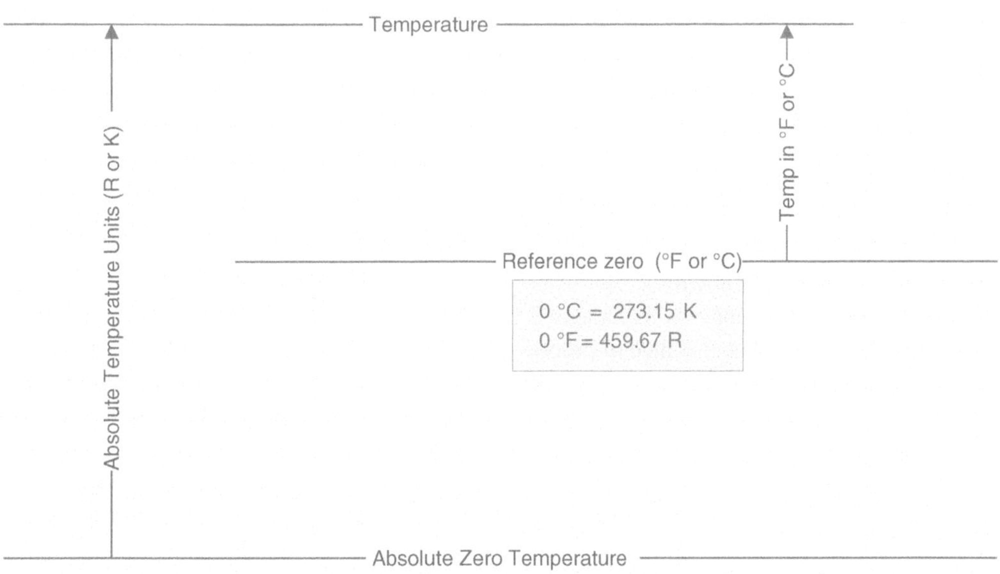

Figure 5.1 Relationship of temperature terms.

zero. Units of Fahrenheit (F) and Celsius (C) have a zero value above absolute zero, so they can have negative values.

Table 5.1 Temperature units of measure.

SI Standard Unit of Measure	
Kelvin	K (*Note:* no degree symbol)
Other Common Units of Measure	
Fahrenheit	°F
Rankin	R
Celsius	°C

5.3 Temperature Instruments

Different types of sensors are used to measure temperature. Electrical sensors convert temperature into an electrical property, such as voltage or resistance, that can be measured with electronic circuitry. Mechanical sensors convert the temperature into a mechanical movement to actuate some devices, such as a valve or indicator, directly.

This section summarizes the most common sensor types used for process measurement (thermocouples and resistance temperature detectors [RTDs]). To be able to choose the right calibration equipment, the calibration technician should understand how each type of sensor works.

5.3.1 Thermocouples

Thermocouples have two wires of dissimilar metals joined at both ends. A current is generated when the temperature at one junction is different from the temperature at the other junction (called the *Seebeck effect*). Measuring the voltage in the circuit is how the temperature is measured.

One end is called the "hot" junction, and the other end is the "cold" junction. The hot junction (measuring junction) is placed in a sheath and used as a measuring element. The cold junction (reference junction) is terminated outside the process at a place where the temperature is known and the voltage is measured. The voltage is proportional to the temperature difference between the two junctions, so to calculate the temperature at the hot junction, the temperature at the cold junction must be known.

Thermocouple terminology is represented in Figure 5.2. An example of an industrial thermocouple that would be installed in process equipment is shown in Figure 5.3.

There are several types of thermocouples; the type is based on the metals that make up the thermocouple. For example, a type T thermocouple is made from copper and constantan wires, whereas type K thermocouples are made from chromel and alumel. Each type is color-coded, and per US standards, the red lead is always negative. Because of the many thermocouple types, a technician must verify the type used and ensure the correct reference table and/or test equipment setup is used for that type.

Thermocouples are usually connected to the control system using a special thermocouple wire made from the same materials as the thermocouple leads. If regular instrument wire is

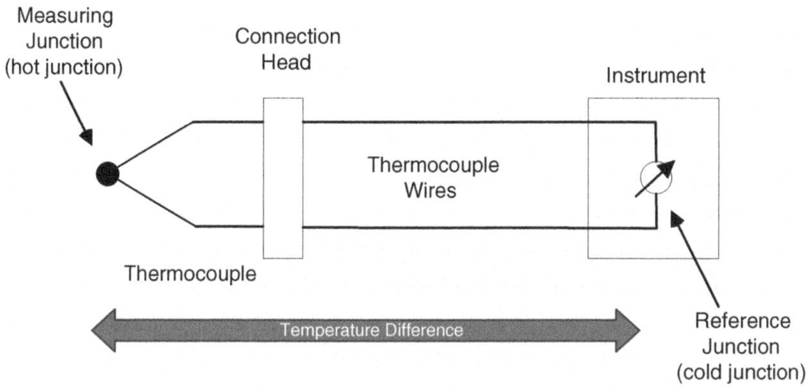

Figure 5.2 Thermocouple principle.

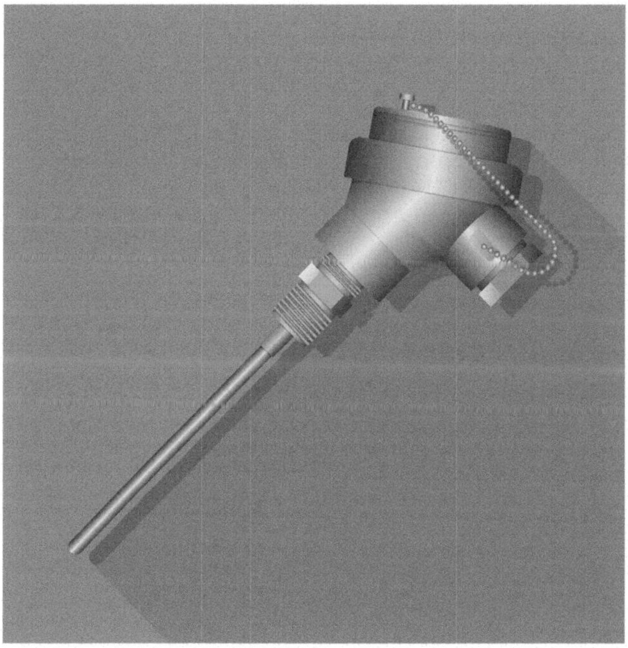

Figure 5.3 Industrial thermocouple. Pepermpron. Adobe Stock Photos.

used, then two more reference junctions are created where the thermocouple leads connect to the wires made of different metals.

Most thermocouple transmitters, calibrators, and control system input cards have cold junction compensation built in to measure the temperature at the cold junction. The cold junction temperature is measured to calculate the hot junction temperature based on the voltage.

5.3.2 RTDs

RTDs are made from a resistance element with a resistance that changes as the temperature changes (Figure 5.4). All RTDs have a positive temperature coefficient, which means the resistance increases as the temperature increases (Figure 5.5). The RTDs most widely used in process

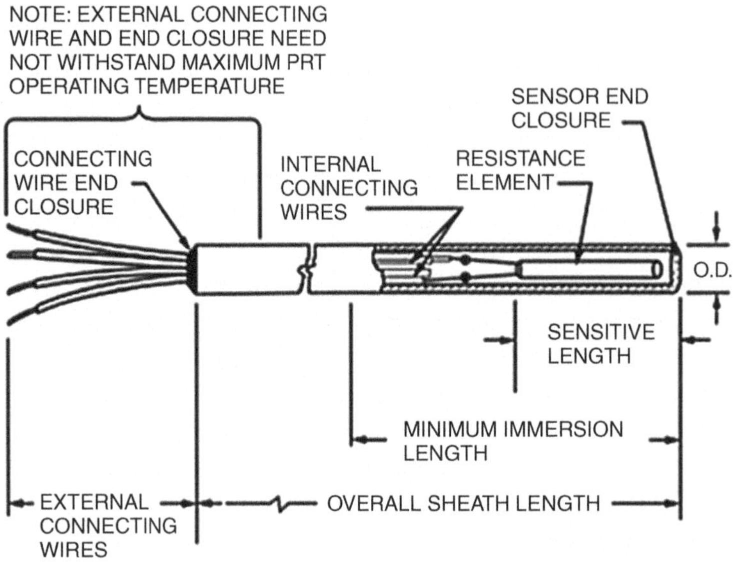

Figure 5.4 Typical industrial platinum resistance thermometer. O.D., outside diameter [2]. Reprinted, with permission, from ASTM E1137/E1137M-08(2020), *Standard Specification for Industrial Platinum Resistance Thermometers*, copyright ASTM International. (ASTM E1137/E1137M-08, 2020).

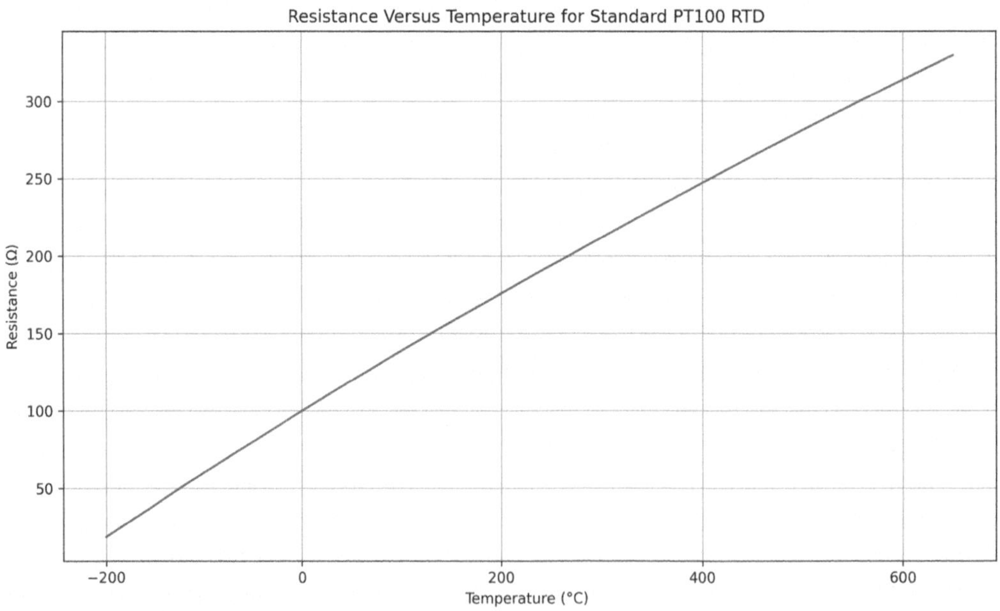

Figure 5.5 Resistance versus temperature for a standard platinum resistance temperature detector (RTD; PT100).

measurement are made from platinum and have a resistance of 100 Ω at 0 °C with a temperature, or alpha, coefficient (α) of 0.00385.

The alpha coefficient (α) is determined from the average slope of the line between 0 °C and 100 °C and the resistance of the RTD at 0 °C (Equation 5.1). The alpha is expressed in ohm (Ω) per ohm per degree Celsius units, where the second ohm refers to the resistance at 0 °C.

$$\alpha\left(\frac{\Omega}{\Omega\,°\mathrm{C}}\right) = \frac{R_{100} - R_0}{100\,°\mathrm{C} \times R_0} \tag{5.1}$$

For the platinum RTD, for example:

$R_0 = 100\,\Omega$

$R_{100} = 138.50\,\Omega$

$\alpha = 0.003850\,\Omega/\Omega\,°\mathrm{C}$

The change in ohms for 100 °C is 38.50 Ω. The change for 1 degree is 0.385 Ω. These are commonly called *PT100 RTDs*.

The resistance output for an RTD is not perfectly linear, and the ohms for a given temperature cannot be determined from the alpha coefficient alone. The complete equation for the RTD output also includes other coefficients that define the nonlinearity. RTD tables are used to determine the temperature given the resistance. The resistance versus temperature relationship for PT100 RTDs is defined in IEC 60751, *Industrial Platinum Resistance Thermometers and Platinum Temperature Sensors* [1].

RTDs are purchased as two-, three-, or four-wire RTDs. The most important thing for a technician to understand is that the RTDs must be connected correctly during initial installation and upon completing calibration. In addition, the technician must be careful to use the correct RTD table as a reference when performing calibrations.

5.3.3 Thermistors

Thermistors are ceramic semiconductors that change resistance when the temperature changes. Thermistors are much more temperature-sensitive than thermocouples or RTDs. This enables thermistors to detect very small changes in temperature that would not be detected by RTDs or thermocouples. Most thermistors have a negative temperature coefficient, which means the resistance decreases as the temperature increases.

However, thermistors are nonlinear and must be used over a small range to provide a linear response. Thermistors are also susceptible to drift and are quite fragile.

5.3.4 Thermometers – Filled-Bulb and Bimetal

Thermometers such as those in Figure 5.6 are used to provide a local indication of temperature. These are typically mechanical devices. Some thermometers also incorporate electrical switches.

Filled-bulb sensors are a closed system filled with a liquid that expands as the temperature increases and contracts when the temperature decreases, producing a proportional change in temperature indication. These are typically used in temperature indicators (thermometers), temperature switches, and mechanical or pneumatic temperature controllers.

Bimetal sensors have two dissimilar metal strips bonded together, typically in a helical spring. The metal strips expand by different amounts as the temperature changes, causing movement. These are typically used for temperature indicators (thermometers) or temperature switches.

Figure 5.6 Thermometer. Aleksandr Ugorenkov. Adobe Stock Photos.

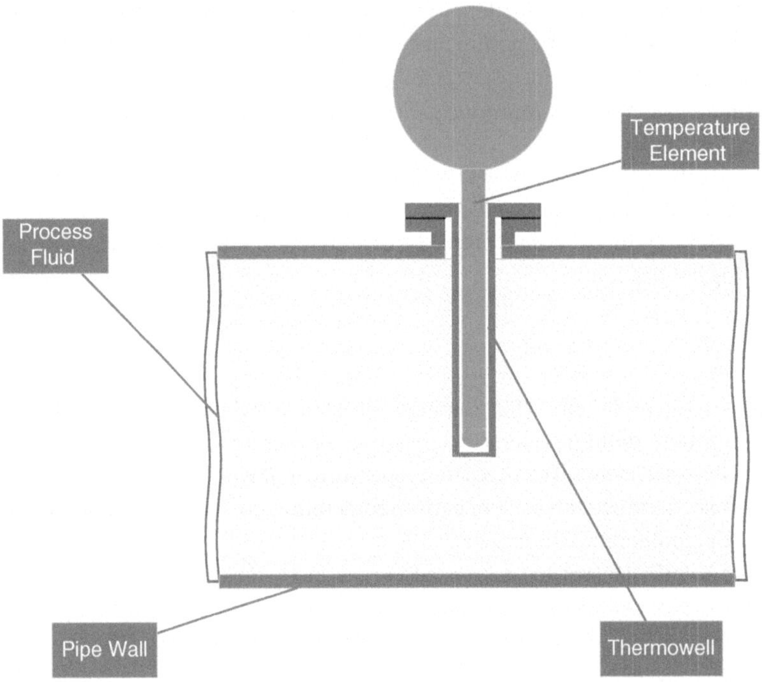

Figure 5.7 Thermowell.

5.3.5 Thermowells

Thermowells are fittings that shield temperature sensors such as thermocouples, RTDs, and thermometers from the process medium. A thermowell is a closed-end tube inserted into the process, with the temperature sensor mounted inside (Figure 5.7). A thermowell allows the control system technician (CST) to remove the sensor for calibration without being exposed to the process and without shutting down the process. If there is no thermowell, the pipe or vessel must be depressurized and emptied to remove the sensor for calibration.

5.3.6 Signal Conversion (Transmitters and Input Cards)

Signal conversion is used to convert the voltage of a thermocouple or the resistance of an RTD/thermistor to a signal that is usable by an instrument system. For example, a temperature transmitter may convert the RTD resistance to a 4–20 mA signal for a control system input. A separate signal conversion device is not always used because many devices (controllers, indicators, and control system inputs) can be configured for direct temperature sensor input.

This chapter focuses on electronic devices, but transmitters and controllers can also be pneumatic. A pneumatic transmitter converts the temperature to a pneumatic signal, such as 3–15 psig, which can be used by a pneumatic controller or chart recorder.

Most transmitters and input cards have the thermocouple and RTD equations built in and will output a signal proportional to the measured temperature. Thermocouple transmitters and input cards also have cold junction compensation built in, so no extra calculations are required in the control system logic. Most calibrators also have cold junction compensation built in, so it is not usually necessary to manually calculate the measured temperature based on the reference temperature.

5.4 Temperature Measurement Uncertainty

CST Level III

What contributes to the uncertainty of a temperature measurement loop? For the experienced technician organizing a calibration program, this is important to know when assessing loop tolerance. Technicians performing calibration and troubleshooting should also understand the contribution each device can make to the overall error.

Figure 5.8 is an example of a temperature measurement loop using an RTD and transmitter connected to a programmable logic controller (PLC) input card. This example is used throughout this chapter.

The RTD, the transmitter, and the input card each have uncertainty. Recall from Section 1.2.8 that the total probable error (TPE) of the reading seen on the operator interface can be calculated using the root sum squared (RSS) equation for the combination of device errors (Equation 5.2).

$$\text{indication error} = \sqrt{(\text{RTD error})^2 + (\text{transmitter error})^2 + (\text{input error})^2} \tag{5.2}$$

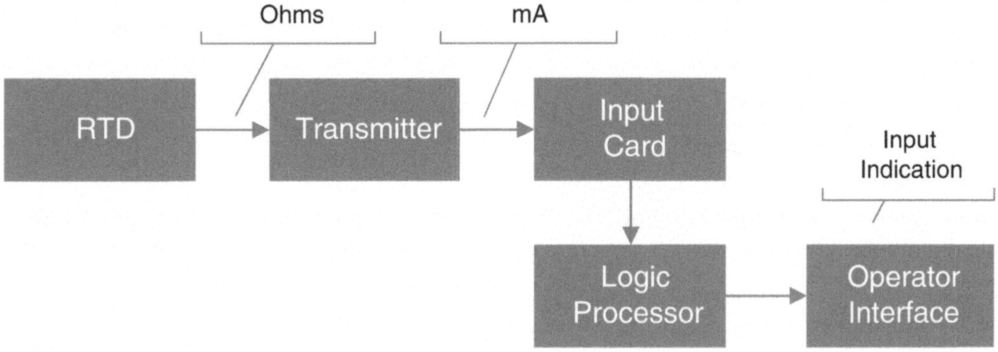

Figure 5.8 Temperature measurement loop. RTD, resistance temperature detector.

The uncertainties associated with each device can be found in the manufacturer's literature. It is usually necessary to use the RSS equation to combine multiple sources of error for each device to determine the likely error associated with each one.

Here is an example of calculating the uncertainty for the loop in Figure 5.8. The conditions for this example are:

- Temperature element: Class B PT100 RTD
- 4–20 mA transmitter output to PLC analog input card
- Transmitter span: 0 °C to 200 °C
- Maximum ambient temperature: 40 °C
- Maximum expected process temperature: 150 °C
- Last calibration for all devices: less than 5 years
- Input card accuracy: 0.1% span
- Specified loop tolerance: 2 °C

5.4.1 RTD Error

RTDs fall into accuracy classes per IEC 60751, *Industrial Platinum Resistance Thermometer Sensors* [1], or ASTM E1137/E1137M-08(2020), *Standard Specification for Industrial Platinum Resistance Thermometers* [2]. Manufacturers will make their devices to meet the requirements of their classification, but it is best to use the manufacturer's data sheet to determine the accuracy. RTDs have some hysteresis, but it is usually negligible. The accuracy of an RTD is a function of temperature (t). Equation 5.3 is an example of the formula for IEC 60751 Class B RTD error tolerance.

$$\pm(0.3 + 0.005|t|)^{\circ}\mathrm{C} \tag{5.3}$$

For example, at $t = 200$ °C, the accuracy is ± 1.3 °C. Tolerances for this Class B RTD over the span of 0 °C to 200 °C are listed in Table 5.2.

Table 5.2 Class B RTD tolerances.

Temperature	Tolerance
0 °C	0.30 °C
50 °C	0.55 °C
100 °C	0.80 °C
150 °C	1.05 °C
200 °C	1.30 °C

RTD, resistance temperature detector.

For calculating the loop uncertainty, use the accuracy calculated at the maximum operating temperature the device will see.

5.4.2 Transmitter

Multiple error components can contribute to the overall accuracy of transmitters. These include the accuracy of the electronics, the resolution of the digital-to-analog conversion, and the effect of ambient temperature on the electronics.

Table 5.3 Example manufacturer data for a temperature transmitter with RTD input.

Input digital accuracy	0.1 °C for an RTD input	0.1 °C
Output digital accuracy	0.02% of span	0.04 °C
Input ambient temperature effect (t = ambient temperature)	$0.0015 \frac{°C}{°C} \times (t - 20\,°C)$	0.03 °C
Output ambient temperature effect (t = ambient temperature)	$0.001 \frac{\%}{°C} \times (t - 20\,°C) \times$ (temperature span)	0.04 °C
Stability	0.25 °C for 5 years	—
TPE for transmitter	$\sqrt{.1^2 + .04^2 + .03^2 + .04^2}$	0.12 °C

RTD, resistance temperature detector; TPE, total probable error.

Table 5.3 is an example of accuracy data for a typical temperature transmitter with an RTD input. (This is just example data; use actual data from the manufacturer's data sheet.) The manufacturer's published numbers are valid within a stated period of stability. To account for drift over time, manufacturers sometimes give a stability number.

5.4.3 PLC Input Card

The input card accepts the 4–20 mA signal from the transmitter. Input cards have some errors associated with converting the digital signal to analog and can also have errors associated with ambient temperature effects on the electronics. The error will be published in the manufacturer's literature. Manufacturers will usually give an overall accuracy that the device can meet within its operating temperature range. For this example, use 0.1% of span (0.1% of 16 mA), which is typical for these devices. This would be 0.2 °C.

5.4.4 Calculating Loop TPE/Loop Tolerance

It is necessary to calculate the TPE for the entire loop to verify that the loop can meet the required loop tolerance. If the loop tolerance is not specified, then TPE can be used as the loop tolerance for the calibration. If a loop calibration shows that the actual loop error is outside the TPE, then the technician should follow the organization's procedure for maintenance or troubleshooting to identify and correct the source of the error. This might involve adjusting or replacing a device.

Use the RSS equation to calculate the TPE for the entire loop (Equation 5.4).

$$\begin{aligned} \text{TPE} &= \pm\sqrt{\text{RTD}^2 + \text{transmitter}^2 + \text{input card}^2} \\ &= \pm\sqrt{1.3^2 + 0.1^2 + 0.04^2 + 0.03^2 + 0.04^2 + 0.2^2} = \pm 1.32\,°\text{C} \end{aligned} \tag{5.4}$$

For this example, the RTD error of 1.3 °C is calculated at 200 °C using Equation 5.3. Because the RTD tolerance changes with the temperature, the TPE of the loop will also change with the temperature.

Corrective action is needed if any loop calibration reading has an error greater than the TPE.

Note that if any of the specifications were relative to reading instead of span, this would be calculated for each calibration temperature instead of just the span temperature.

5.5 Temperature Calibration

5.5.1 Introduction

There are two common methods for calibrating temperature devices such as thermometers, thermocouples, RTDs, thermistors, transmitters, and control system input cards. Method 2 is preferred because it is a complete loop calibration that includes all the devices in the loop (e.g., sensor, transmitter, and input card). The methods are:

1) Simulate a sensor output to calibrate just a transmitter or input card.
2) Physically heat or cool the sensor to calibrate the sensor and the entire loop.

Simulators are used to calibrate a transmitter or control system input card. The simulator will generate a voltage to simulate a thermocouple or a resistance to simulate a thermistor or RTD. Most simulators can simulate multiple types of sensors; the type is selectable (e.g., Type K thermocouple or PT100 RTD).

5.5.2 Temperature Calibration Equipment

"To calibrate a temperature sensor, it must be inserted into a known temperature and compared against a precision calibration standard sensor" ([3], p. 100). Temperature calibrators that produce physical heat or cooling include dry-block (or dry-well) and temperature calibration baths.

In a dry-block calibrator, a metal block is heated or cooled to a set temperature. The sensor is inserted into the block. The output of the sensor under test is compared with the output of the standard reference sensor, which can be internal to the calibrator or an external device. Portable dry-block calibrators, such as the Beamex MC6-T in Figure 5.9 [4], can be used in the field or a shop to heat or cool a sensor to a specified temperature. For example, a PT100 RTD chilled to 0 °F should have a resistance of 100 Ω.

A metrology well is a very high-accuracy, dry-block calibrator that provides better heat distribution in the well than a standard dry-block.

Temperature bath calibrators use a stirred fluid in a tank to heat or cool a sensor. Temperature baths with external sensors are more accurate than dry-blocks, so they can be used where lower uncertainty is required. Temperature baths are usually slower to change the temperature than dry-blocks and are mostly used on the bench in a shop or lab.

These devices have internal sensors to control the temperature. Some can accommodate an external reference probe to improve accuracy (Figure 5.10).

Figure 5.11 shows the components of a typical dry-block calibrator. Dry-blocks use inserts for the heated/cooled cavity that are bored out to various diameters to accommodate a wide variety of sensor diameters. These inserts may be easily interchanged as may be required. It is important to make sure that the insert is the appropriate length and diameter for the sensor that is being calibrated.

Sources of measurement uncertainty for a dry-block calibrator are generally:

- Stated accuracy of the reference measurement.
- Stability (how well the calibrator holds the temperature).
- Axial homogeneity (how consistent the temperature is along the length of the sensor in the insert).
- Radial homogeneity (how consistent the temperature is horizontally between the sensors).
- The loading effect (multiple sensors in the block).

Figure 5.9 Portable dry-block calibrator [4]. Reproduced with permission of Beamex.

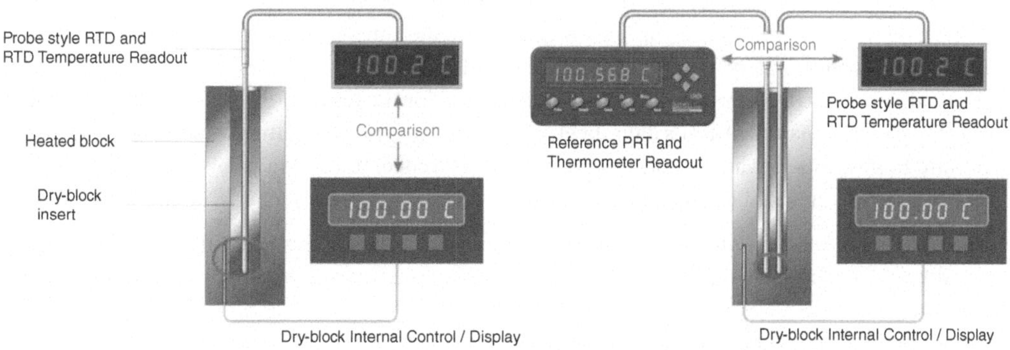

Figure 5.10 Internal and external reference probes in a dry-block calibrator. PRT, platinum resistance thermometer; RTD, resistance temperature detector [5]. Reproduced with permission of Fluke Corporation.

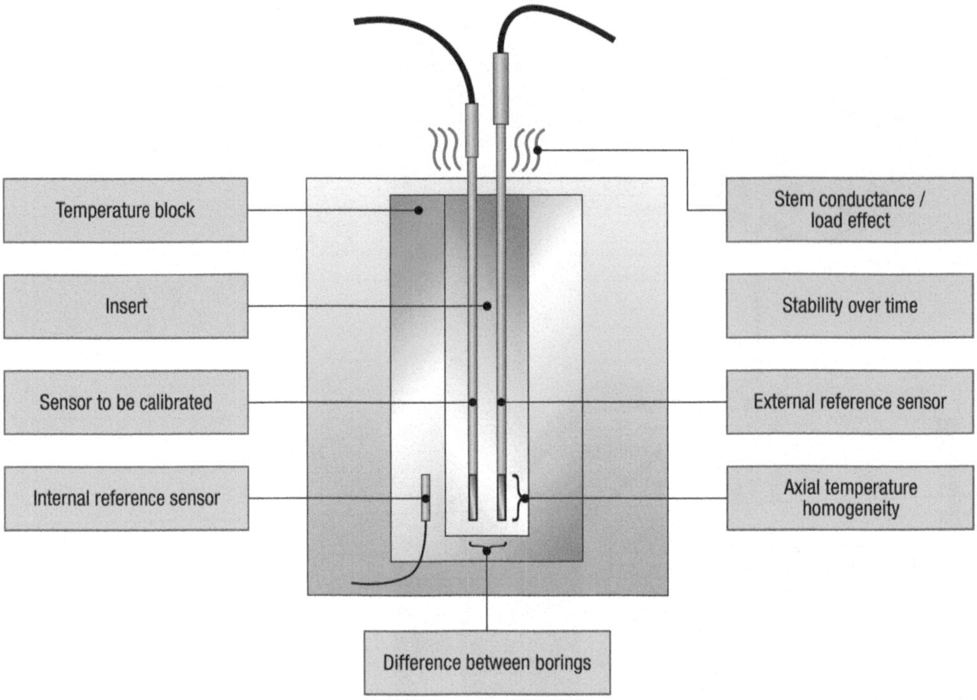

Figure 5.11 Dry-block calibrator components [6]. Reproduced with permission of Beamex.

Some important things to consider when using a dry-block calibrator:

- It is best if the sensor reaches the bottom or near the bottom of the block. This is the region where the manufacturer's accuracy is stated.
- The internal reference sensor is usually mounted inside the block, so it measures the temperature of the block, which can be different than the temperature of the insert. Using an external reference sensor can improve accuracy.
- When using an external reference sensor, it should be inserted to the same depth as the sensor being calibrated to minimize error. This way, they are measuring at the same depth in the block. The temperature can vary along the length of the block.
- After adjusting the temperature set point, it takes time for the temperature of the system to stabilize. Consult the manufacturer's data sheet for the time to allow for getting a stable reading. Also, be sure to give time for all sensors, including the reference, to reach a stable reading. Different types of sensors have different response times.

A portable dry-block multifunction calibrator will be used for the examples in this chapter. The calibrator in these examples is a Fluke 9102S using the internal reference (a data sheet is in Appendix E).

Equation 5.5 is the combined uncertainty calculation for the 9102S.

$$\text{uncertainty} = \sqrt{\text{accuracy}^2 + \text{stability}^2} = \sqrt{0.25^2 + 0.05^2} = \pm 0.25\,^\circ\text{C} \tag{5.5}$$

5.5.3 Safety

If the sensor is in a thermowell, it can be safely removed for calibration. Be sure to unscrew the sensor from the thermowell; do not unscrew the thermowell from the pipe.

If the sensor is not in a thermowell, the piping would have to be isolated and depressurized, and any hazards associated with the process fluid would have to be considered before removing the sensor.

5.5.4 Temperature Calibration Examples

5.5.4.1 Dial Thermometer Example

A dial thermometer consists of a sensing probe connected to a dial indicator (Figure 5.12). The temperature causes the dial pointer to move across a graduated scale. The graduated scale provides an indication over a range of temperatures. Specifications are given in Table 5.4.

The goal of the calibration is to verify that the pointer "points" to the correct value over the range of the instrument.

The test equipment will be the Fluke 9102S handheld, portable dry-block calibrator with an internal reference sensor (described in Section 5.5.2).

Verify the test uncertainty ratio (TUR): The combined uncertainty using the internal reference is ±0.25 °C (Equation 5.5). The gauge's required tolerance is ±5 °C, so the TUR is better than 4:1. A separate temperature standard could be used to obtain more accuracy, but the internal reference is sufficient for the specified tolerance.

Figure 5.12 Dial thermometer. Bart. Adobe Stock Photos.

Table 5.4 Example dial thermometer specifications.

	Thermometer
Tag	TG-100
Range	0 °C to 120 °C
Tolerance	±5 °C
Test points	0%, 25%, 50%, 75%, 100% 0 °C, 30 °C, 60 °C, 90 °C, 120 °C

To calibrate the thermometer, it must be removed from the system and inserted into the calibrator.

If the thermometer is in a thermowell, it can be safely removed for calibration. Be sure to unscrew the thermometer from the thermowell; do not unscrew the thermowell from the pipe. If the thermometer is not in a thermowell, the piping would have to be isolated and depressurized, and any hazards associated with the process fluid would have to be considered before the thermometer is removed.

Select a sleeve that fits the thermometer probe, and install it in the dry-block. Insert the thermometer into the sleeve.

Start at the minimum test point of 0 °C. Once the readings have stabilized, record the as-found readings for the temperature standard and the unit under test.

The user manual for this calibrator gives the amount of time for complete stabilization as 7 minutes. The heating time is 10 minutes. Advanced electronic calibrators will generally indicate that the reference has stabilized.

No adjustments are made until all as-found readings are recorded for all test points. For the gauge in this example, each mark is 1 degree, so it is impossible to read the temperature better than half-degree increments.

Repeat the process, and record the as-found readings for the thermometer and the standard at the 30 °C, 60 °C, 90 °C, and 120 °C set points.

Repeat this process from 120 °C back down to 0 °C to check for hysteresis.

Now that the as-found data has been collected, it can be evaluated. The as-found results are shown in Tables 5.5 and 5.6.

Table 5.5 Thermometer calibration data (increasing).

			As-Found		
Input %	**Applied Input (Standard)**	**Desired Indication**	**Increasing Indication**	**Error in Engineering Units**	**Error in %**
0	0.120 °C	0 °C	1 °C	0.88 °C	
25	30.080 °C	30 °C	3 °C	0.92 °C	
50	60.170 °C	60 °C	62 °C	1.83 °C	
75	89.990 °C	90 °C	92 °C	2.01 °C	
100	120.020 °C	120 °C	122 °C	1.98 °C	

Table 5.6 Thermometer calibration data (decreasing).

			As-Found		
Input %	**Applied Input (Standard)**	**Desired Indication**	**Decreasing Indication**	**Error in Engineering Units**	**Error in %**
100	120.020 °C	120 °C	122 °C	1.98 °C	
75	90.050 °C	90 °C	92 °C	1.95 °C	
50	59.980 °C	60 °C	59 °C	−0.98 °C	
25	30.110 °C	30 °C	29 °C	−1.11 °C	
0	0.050 °C	0 °C	2 °C	1.95 °C	

The gauge tolerance is ±5 °C, so this gauge is acceptable. If it were not, the gauge would have to be adjusted or replaced.

5.5.4.2 Temperature Loop Example

The Loop Consider the temperature loop example from Section 5.4. For this, a temperature measurement is indicated on the operator interface in the control room and is also used in a temperature control loop with a normal set point of 75 °C (Figures 5.13 and 5.14). The human-machine interface (HMI) is configured for a low alarm of 75 °C and a high alarm of 85 °C. These alarms can be verified during the loop calibration.

The uncertainty tolerance of the input measurement given by the engineering department is ±1.5 °C to maintain product quality. In Section 5.4, the TPE for the loop was calculated (Equation 5.4) to be no worse than 1.32 °C within the transmitter's span, so the equipment in the loop is capable of meeting the tolerance specified.

The Class B PT100 RTD is connected to a transmitter that is configured to output 4 mA to 20 mA, representing 0 °C to 200 °C. The current signal is connected to an analog input card on a PLC. The HMI display is configured to indicate the temperature from 0.0 °C to 200.0 °C.

Equipment The test procedure task list (Appendix D, Temperature Loop Calibration Using a Dry-Block Calibrator) calls for calibrating the loop by inserting the RTD into a portable dry-block calibrator and executing a 5-point test over the configured range of 0 °C to 200 °C, first increasing and then decreasing.

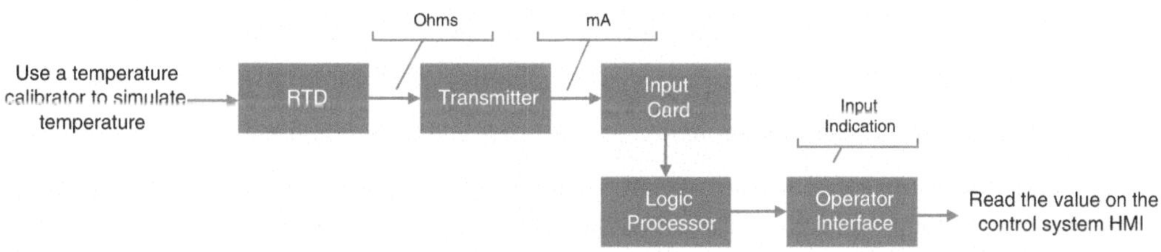

Figure 5.13 Temperature loop example. HMI, human-machine interface; RTD, resistance temperature detector.

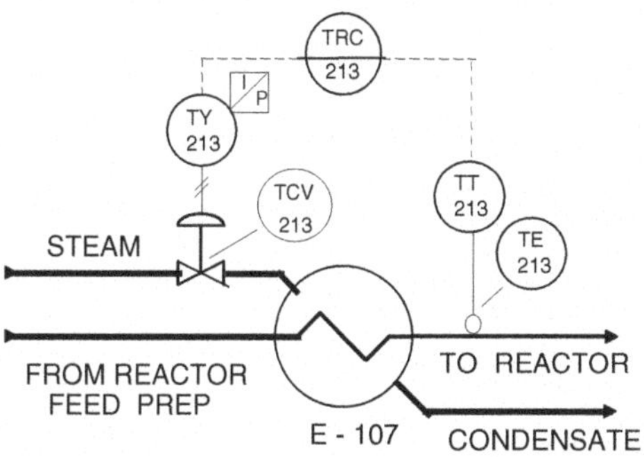

Figure 5.14 Temperature loop 213 for the example. I/P, current-to-pressure; TCV, temperature control valve; TE, temperature element; TRC, temperature recording controller; TT, temperature transmitter; TY, temperature auxiliary device.

Ensure that the TUR is 4:1 for the calibration equipment total uncertainty compared with the required loop tolerance. For this example, the dry-block uncertainty is ±0.25 °C (Equation 5.5), and the loop tolerance is ±1.5 °C; the TUR of 6:1 is acceptable.

Preparation The procedure also says that the system must be in a state that will allow the safe performance of the calibration. The system controlled by this controller should be shut down, or the controller should be placed in manual operation. If the controller remained in automatic during the calibration, the simulated signals would cause output responses that could be detrimental to the equipment and process. The CST must communicate with operations so they can put the loop in a state to accommodate the calibration.

Calibrating The RTD will be removed from the thermowell and inserted into the dry-block calibrator. The calibration will test five points increasing in temperature. The procedure does not call for checking hysteresis by running the calibrations with decreasing temperature.

For each temperature step:

1) Adjust the calibrator set point as shown in the Desired Indication column (Figure 5.15).
2) Record the set point in the Desired Indication column.
3) Wait until the calibrator indicates that the temperature is stabilized.
4) Record the temperature indicated on the calibrator in the Applied Input column. This temperature is measured by the dry-block's internal reference sensor, so it might vary slightly from the set point.
5) Record the reading on the operator interface computer HMI in the Increasing Indication column.
6) Calculate the error for each step and record it in the Error in the EU column (Equation 5.6). The indication is read from the operator interface. The applied input is read from the dry-block; it is the temperature measured by the dry-block's internal reference sensor.

$$\text{error} = \text{indication} - \text{applied input} \tag{5.6}$$

The transmitter output could also be measured and recorded. The reason for measuring the transmitter output for a loop calibration is to help determine which device to adjust if an adjustment is necessary.

Because the tolerance is given in Celsius, it is not necessary to calculate the error in percentage at each step. If the tolerance were given in percent of reading, then %error would be calculated

Input %	Applied Input (Standard)	Desired Indication	Increasing Indication	As-Found	
				Error in Engineering Units	Error in %
0	0.01 °C	0 °C	0 °C	–0.01 °C	
25	50.10 °C	50 °C	49.6 °C	–0.50 °C	
50	99.98 °C	100 °C	99.1 °C	–0.88 °C	
75	150.15 °C	150 °C	148.7 °C	–1.45 °C	
100	200.91 °C	200 °C	198.2 °C	–2.71 °C	

Figure 5.15 As-found data for the temperature loop calibration example.

with Equation 5.7. If tolerance were in percentage of span, then %error would be calculated with Equation 5.8.

$$\%\text{reading error} = \frac{\text{error}}{\text{applied input}} \tag{5.7}$$

$$\%\text{span error} = \frac{\text{error}}{\text{URV} - \text{LRV}} \tag{5.8}$$

Evaluate the Results The specified tolerance for this loop is ±1.5 °C, and there are two readings with errors greater than or equal to 1.5 °C, so this loop will require troubleshooting and/or adjustment.

The error could come from any of the components in the loop: RTD, transmitter, input card, PLC program, or HMI configuration. Assuming the PLC program and HMI configuration are correct, the CST can calibrate the RTD, transmitter, and input card. The only one of these that is adjustable in the field is the transmitter. The RTD or input card must be replaced if it fails individual calibration.

5.5.4.3 RTD Calibration Check Example

Although temperature sensors are usually checked as part of calibrating the connected loop, it may be necessary to check a suspect temperature sensor or to verify the accuracy of a new device.

The principles of this RTD calibration check would apply to any sensor type. The only exception would be a thermocouple calibration check if the test equipment does not have automatic cold junction compensation. Most calibration test equipment performs temperature compensation, but it is important to verify that it has this capability. If the calibrator does not have thermocouple temperature compensation, a second temperature standard must be used to measure the reference junction ambient temperature and compensate manually.

For this example, the RTD from the previous example (Section 5.5.4) will be checked at the same five points from 0 °C to 200 °C using the RTD calibration task list in the procedure. The same dry-block calibrator will be used to simulate the temperature on the RTD, which is still inserted in the calibrator. Resistance will be measured using the multifunction calibrator or a multimeter (Figure 5.16).

Using Equation 5.3, the tolerance for the RTD calibration is calculated at each temperature and listed in Table 5.7.

The multifunction calibrator used in this example has an RTD input. It will indicate the temperature represented by the input resistance if the correct sensor type is selected in the configuration. This feature can be used, or the resistance can be measured directly using the calibrator or a calibrated multimeter and compared with the values in the RTD table in Appendix F. Table 5.8 shows both values.

For this example, the RTD measurements are all within tolerance, so the RTD is acceptable. If the RTD indication was incorrect, the RTD would be replaced, and the loop calibration would be performed again. The error contributed by the RTD to the loop is small compared with the loop error in Figure 5.15, so it is likely not the main reason for the loop being out of tolerance.

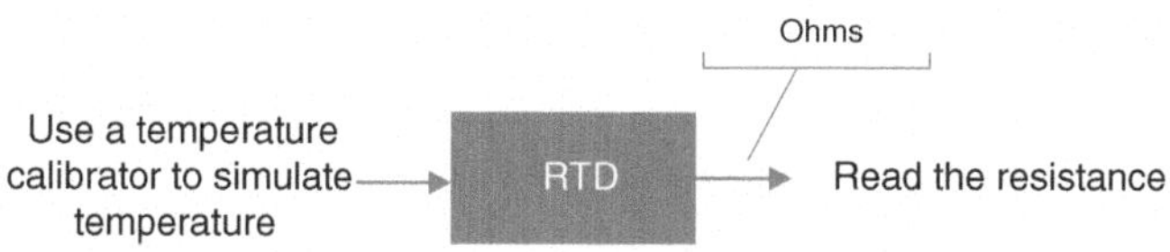

Figure 5.16 Sensor calibration. RTD, resistance temperature detector.

Table 5.7 RTD calibration tolerances.

Temperature	Tolerance
0 °C	0.30 °C
50 °C	0.55 °C
100 °C	0.80 °C
150 °C	1.05 °C
200 °C	1.30 °C

RTD, resistance temperature detector.

Table 5.8 Sensor calibration results.

			As-Found					
Input %	Applied Input (Standard)	Desired Indication	Expected Resistance	Measured Resistance	Calibrator Reading	Reading From Table	Error in Engineering Units	Error in %
0	0.00 °C	0 °C	100.000 Ω	100.10 Ω	0.26 °C		0.1 °C	0.1
25	50.00 °C	50.0 °C	119.397 Ω	119.42 Ω	50.05 °C		0.05 °C	0.1
50	100.00 °C	100.0 °C	138.505 Ω	138.54 Ω	100.10 °C		0.10 °C	0.1
75	150.00 °C	150.0 °C	157.325 Ω	157.38 Ω	150.15 °C		0.15 °C	0.1
100	200.00 °C	200.0 °C	175.856 Ω	175.93 Ω	200.20 °C		0.20 °C	0.1

Notice that the expected resistance values are based on the actual temperature standard reading obtained, not the test point.

5.5.4.4 Temperature Transmitter Example

Because the RTD passed calibration, the next troubleshooting step is to calibrate the transmitter by simulating the RTD input to the transmitter and reading the analog output to see if the transmitter is behaving within tolerance (Figure 5.17).

Disconnect the RTD from the transmitter and disconnect the transmitter from the PLC input. Use a simulator in place of the sensor. The multifunction calibrator in the previous examples can simulate an RTD. If the sensor were a thermocouple, a thermocouple simulator of the correct type would be connected to the transmitter input.

Figure 5.18 shows an example of an RTD simulator connected to a transmitter. This device sources a simulated RTD signal (resistance). It can also power the 4–20 mA loop and read the transmitter output (Figure 5.19).

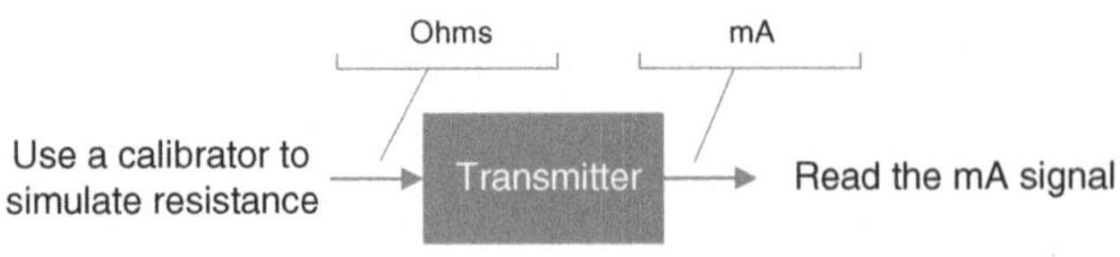

Figure 5.17 Transmitter calibration.

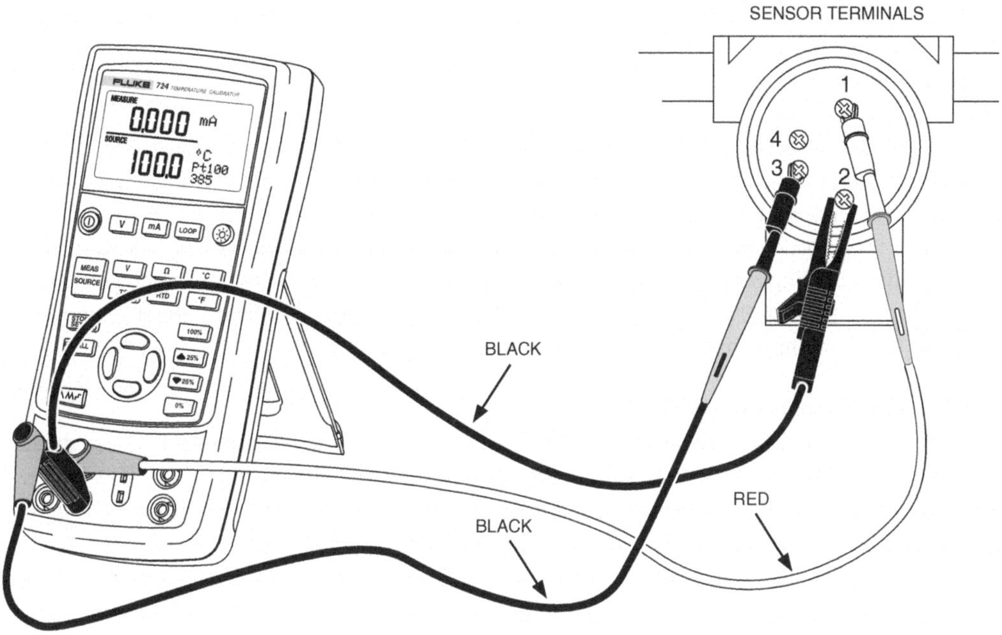

Figure 5.18 Temperature calibrator hookup [7]. Reproduced with permission of Fluke Corporation.

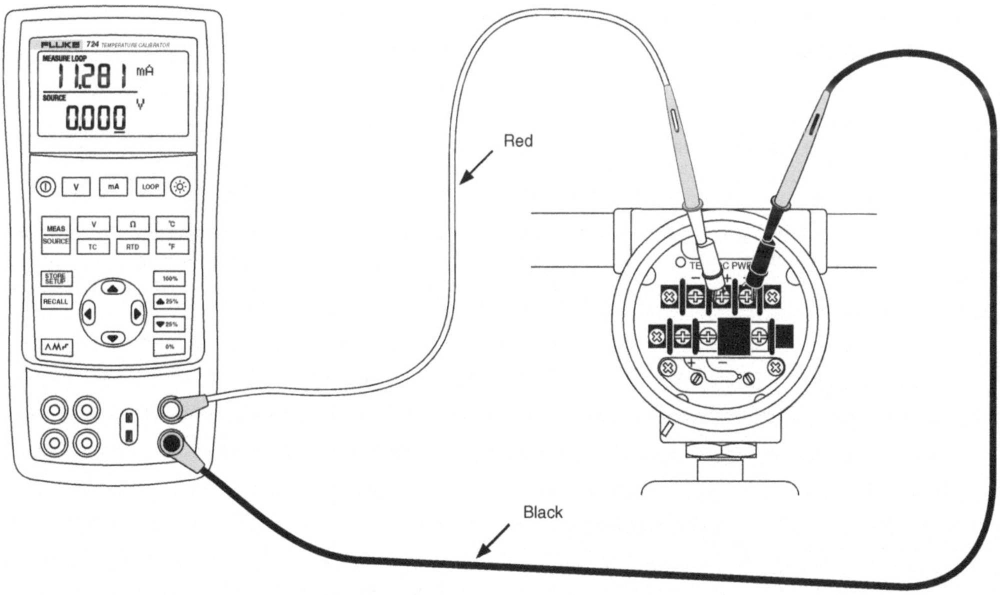

Figure 5.19 Connections for supplying loop power and measuring milliamp (mA) signal [7]. Reproduced with permission of Fluke Corporation.

From Table 5.3 in Section 5.4, the TPE of the transmitter is ±0.12 °C. How much of a change in the loop current represents 0.12 °C?

Transmitter 16 mA span: 0 °C to 200 °C = 4 mA to 20 mA

±0.12 °C = 0.0096 mA

Given the actual transmitter calibration results (Table 5.9), does the transmitter require calibration? The error at some test points is greater than the TPE of the transmitter, so the transmitter requires adjustment or replacement.

Why does this procedure require test points of 10%, 50%, and 90% instead of 0%, 50%, and 100%? The transmitter reading may go too far below the minimum or above the maximum to give a valid reading. If the 0% and 100% test points were used, and the reading was below the 0% reading or above the 100% reading, it might not be possible to calculate the error.

5.5.4.5 Analog Input Card Example

Calibration checks are usually performed on input and output cards in a control system during commissioning. If the loop fails calibration, the associated analog input point will be checked as a troubleshooting step. The multifunction calibrator is used to simulate the 4–20 mA analog signal, and the value can be read in the PLC logic, on the HMI, or on both (Figure 5.20).

If the PLC card sources power for the loop, the calibrator must be set to sink, not source.

Once the system is in a safe condition and the controller is placed in manual, disconnect the controller input leads, and connect a milliamp simulator to the input. To perform the calibration, adjust the milliamp simulator to the desired set points and record the controller display.

Most input and output cards do not have adjustments accessible to the user. If the card fails calibration, it should be replaced and returned to the manufacturer. The programming can compensate for calibration, but this should be a temporary measure.

Alarm set points and control output action can also be verified during the calibration. This is not required, and many calibration procedures would not include it. To perform the alarm checks, the

Table 5.9 Transmitter calibration results.

Test Point (%)	Simulated Temperature (°C)	Input Resistance (Ω)	Expected Transmitter Output (mA)	Actual Transmitter Output (mA)	Error (mA)
10	20.0	107.79	5.60	5.50	−0.10
50	100.0	138.51	12.00	11.88	−0.12
90	180.0	168.48	18.40	18.27	−0.13

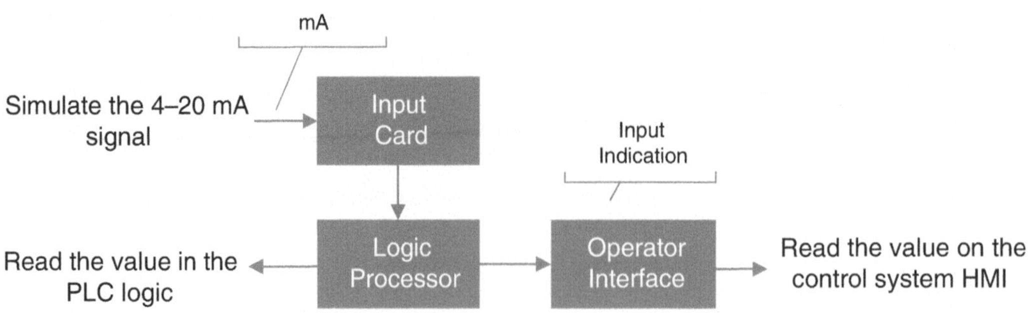

Figure 5.20 Input card calibration. HMI, human-machine interface; PLC, programmable logic controller.

milliamp simulator is adjusted as necessary to determine the temperature at which the low and high alarms trip and reset. To check the controller output for proper operation, adjust the milliamp simulator above and below the set point safely to verify that the proper output action is occurring with the controller in automatic mode.

5.5.4.6 Temperature Switch Example

A temperature switch is a device that senses temperature and changes state at the programmed or adjusted set point. A temperature switch usually has at least one set of normally open (NO) and one set of normally closed (NC) contacts, but it may have only an NC contact, a NO contact, or more than one set of each. NO and NC refer to the state of that contact with the switch de-energized. The tricky part of calibrating a temperature switch is to know whether the switch should trip with increasing or decreasing temperature. Typically, a high-temperature switch would trip with increasing temperature and vice versa.

The switch calibration is checked by placing the sensor in the dry-block and measuring across the applicable contact with a multimeter. The calibrator temperature is adjusted past the set point. The multimeter will read the voltage or resistance change when the switch changes state.

Review Questions

5.1 Two common temperature elements are RTDs and thermocouples. Which device outputs voltage related to temperature, and which changes resistance with temperature?

5.2 Why is it important to know if a temperature element is installed in a thermowell?
a) Thermowells change the properties of a sensor.
b) Thermowells eliminate the need for calibration.
c) If a sensor is in a thermowell, it can be removed for calibration without shutting off the process.
d) The thermowell requires special calibration equipment.

5.3 What is the function of a temperature transmitter?
a) Convert a raw RTD or thermocouple signal to a standard output (such as 4–20 mA) for the control system
b) Control the process temperature based on a set point
c) Protect the sensor from the process fluids
d) Improve the sensor accuracy

5.4 Which components must be considered for calculating a temperature loop TPE?

5.5 Why might a procedure call for using an external reference probe in a dry-block calibrator instead of using the internal reference probe?
a) The external probe eliminates the need for stabilization time.
b) The external probe does not require power.
c) The built-in reference can only be used for certain types of sensors.
d) The external probe provides a more accurate measurement comparison with the device being tested.

5.6 What is the maximum error in degrees Celsius that is expected of an ASTM Grade A RTD at 50 °C? (Use the RTD specification in Appendix C.)

5.7 For the loop calibration example in Section 5.5.4.2, calculate the percentage reading error for each input point listed in Figure 5.15.

5.8 Regarding the example in Section 5.5.4.4, if the sensor was a thermocouple instead of an RTD, what kind of signal would be simulated to the transmitter input?

5.9 Which equipment would ***not*** be used when calibrating a temperature loop or temperature sensor?
a) Digital multimeter
b) Dry-block (dry-well)
c) Temperature bath
d) Hand pump

References

Note: The dates and editions for standards and other references in this chapter are current as of publication. The reader is encouraged to check for later revisions or editions.

1 International Electrotechnical Commission (IEC), IEC 60751: 2022, *Industrial Platinum Resistance Thermometers and Platinum Temperature Sensors*. Geneva, Switzerland: IEC, 2022.

2 American Society for Testing and Materials (ASTM), ASTM E1137/E1137M-08 (2020), *Standard Specification for Industrial Platinum Resistance Thermometers*. West Conshohocken, PA: ASTM International, 2020.

3 Rosemount/Emerson, *The Engineer's Guide to Industrial Temperature Measurement*, 2013th ed. Chanhassen, MN: Rosemount, Inc., 2013.

4 Beamex, *Beamex MC6-T Multifunction Temperature Calibrator and Communicator*. Beamex. https://www.beamex.com/app/uploads/2024/06/mc6-t-brochure.pdf. Accessed December 12, 2024.

5 Fluke, "How to do a temperature sensor comparison calibration." Fluke Corporation. https://www.fluke.com/en/learn/blog/temperature-calibration/how-do-temperature-sensor-comparison-calibration. Accessed January 25, 2025.

6 Laurila, Heikki, 2018, August 23, "Uncertainty Components of a Temperature Calibration Using a Dry Block," Calibration White Paper from *Beamex Blog*. https://blog.beamex.com/uncertainty-components-temperature-calibration-dry-block. Accessed February 9, 2025.

7 Fluke, February, *Fluke 724 Temperature Calibrator Users Manual*, Rev. 1. Everett, WA: Fluke Corporation, 2000.

6

Pressure Instrument Calibration

6.1 What Is Pressure?

Pressure is the force applied to an area, as shown in Figure 6.1. It can be expressed in units of force/area, such as pounds-force per square inch.

$$\text{pressure} = \frac{\text{force}}{\text{area}}$$

A force of 1 pound-force acting on an area of 1 in.^{2} exerts a pressure of 1 lbf/in.^{2} (1 psi).

$$Pressure = \frac{Force}{Area}$$

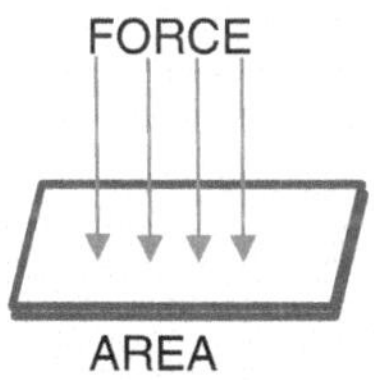

Figure 6.1 Pressure illustrated.

6.2 Pressure Terms and Units of Measure

Frequently used units of measure for pressure are listed in Table 6.1. The following notes provide additional information regarding the pressure terms.

- The units in. H_2O and oz./in.^{2} are frequently used for low-pressure applications.
- The unit ft. H_2O (feet of water head) is frequently used for pump head and level measurement applications.
- Millimeters of mercury (mm Hg) are often seen in medical applications.
- Torr is used in very low-pressure and vacuum applications.
- Some conversion factors for pressure units are included in Appendix F, "Unit Conversions."

Table 6.1 Units of measure for pressure.

SI Standard Unit of Measure	
Pascal	**Pa**
Other Common Units of Measure	
Pound-force per square inch	psi
Pounds-force per square inch absolute	psia
Pounds-force per square inch absolute	psig
Inches of water head	in. H_2O
Feet of water head	ft. H_2O
Ounces per square inch	oz./in.^{2} or oz./sq. in.
Bar	bar
Inches of mercury	in. Hg
Millimeters of mercury	mm Hg
Millimeters of water	mm H_2O
Torr	Torr

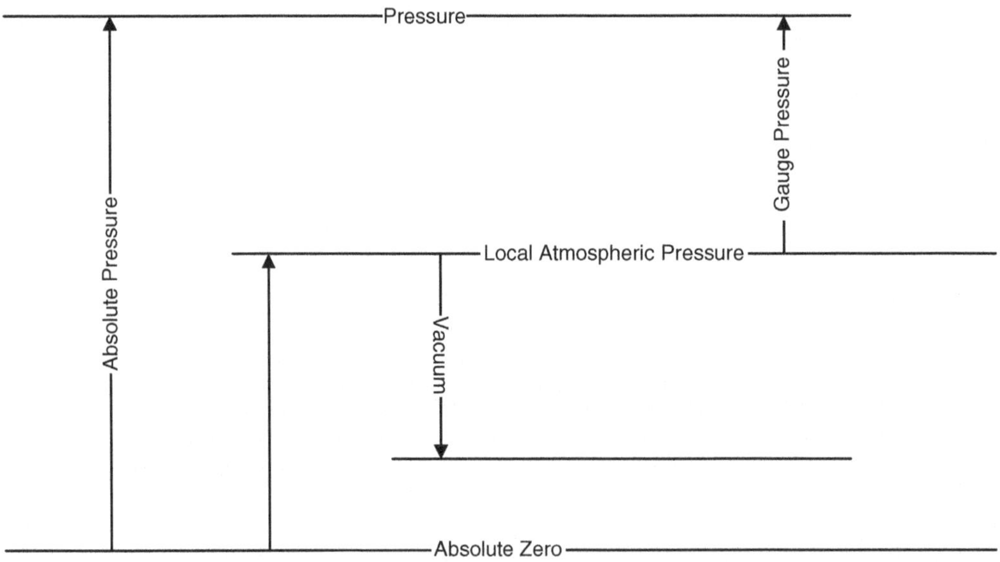

Figure 6.2 Relationship of pressure terms.

Figure 6.2 shows the relationship of important pressure terms.

- Absolute zero pressure is the pressure of a perfect vacuum. There is no material at all in a perfect vacuum.
- Absolute pressure is the amount of pressure above absolute zero. It is given in absolute units such as pounds per square inch absolute (psia).
- Local atmospheric pressure is the pressure exerted by the atmosphere at a location. At sea level, it is 14.7 psia = 1 atm = 1 bar.

Tech Tip

Atmospheric pressure is the force exerted by the weight of tiny air particles (air molecules). Although air molecules are invisible, they still have weight and take up space. Atmospheric pressure is approximately 14.7 lbf/in.^{2} (psia) at sea level. This means that if 1 in.^{2} column of air from the ground to the upper atmosphere could be put on a scale, it would weigh 14.7 pounds. We do not feel this pressure because it also acts internally and is thus balanced.

- Gauge pressure is the pressure relative to local atmospheric pressure.

  gauge = absolute – atmospheric

  Gauge pressure is given in units such as barg or psig. It is important to know if you are dealing with gauge or absolute pressure when calibrating.
- Vacuum pressure is below atmospheric pressure. It is usually given in gauge units (psig), but it will be negative, or there will be an indication that it is a vacuum measurement.
- Differential pressure (DP) is the difference between two pressure measurements. It is usually given without reference to absolute, such as pounds per square inch (psi) or pounds per square inch differential (psid). For example, pressure might be measured at two different points in a pipe (Figure 6.3). Differential pressure is often used in flow measurement (pressure drop across a restriction) or to detect fouling in filters, coolers, and other equipment.

Pressure can be given in terms of the pressure exerted by a column of liquid. This is called *hydrostatic pressure*. In the example in Figure 6.4, the water level in an atmospheric pressure tank is 5 feet above the pressure transmitter. The conversion from feet of water (at 4 °C) to pounds per square inch is shown in Equation 6.1. It is necessary to specify the temperature because the water density changes with temperature. Any standard pressure unit expressed in terms of level must be at a specified temperature for the fluid.

$$2.31 \text{ ft. } H_2O = 1 \text{ psi} \tag{6.1}$$

So, in this example, the pressure can be given as 5 ft. H_2O or 2.16 psig (Equation 6.2).

$$\frac{1 \text{ psi}}{2.31 \text{ ft. } H_2O} \times 5 \text{ ft. } H_2O = 2.16 \text{ psi} \tag{6.2}$$

The density or specific gravity of the liquid must be known. Water density changes with temperature changes, as do most other liquids. The conversion given is for water at 4 °C. At a different temperature, the pressure would be adjusted for the specific gravity at that temperature.

If the liquid was oil with a specific gravity of 0.8, then the pressure would be (Equation 6.3):

$$P = 5 \text{ ft.} \times 0.8 = 4 \text{ ft. } H_2O \tag{6.3}$$

Similar units are inches of water, inches of mercury, and millimeters of mercury. These are typically used in low-pressure and differential pressure measurements.

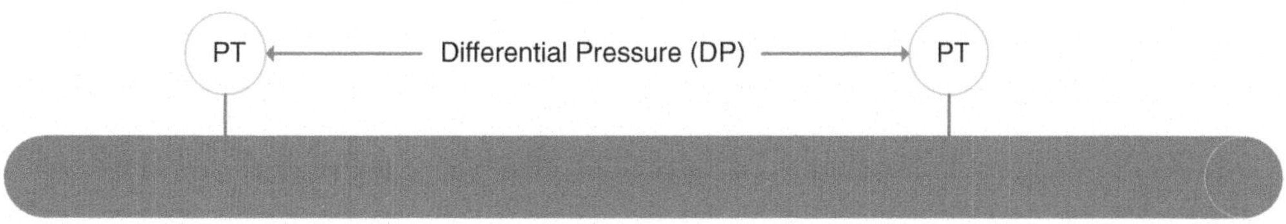

Figure 6.3 Differential pressure. PT, pressure transmitter.

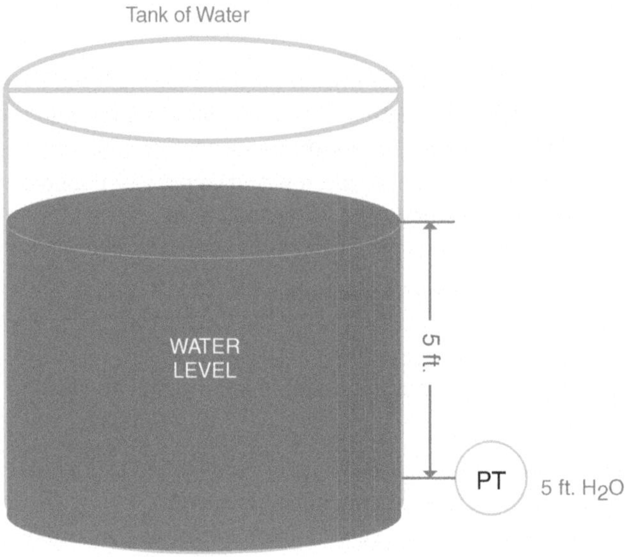

Figure 6.4 Hydrostatic pressure. PT, pressure transmitter.

6.3 Types of Pressure Instruments

6.3.1 Pressure Gauges

Pressure gauges are often used as local indicators of process pressure. Pressure changes applied to the gauge cause the elastic element to expand and contract. The element movement is translated

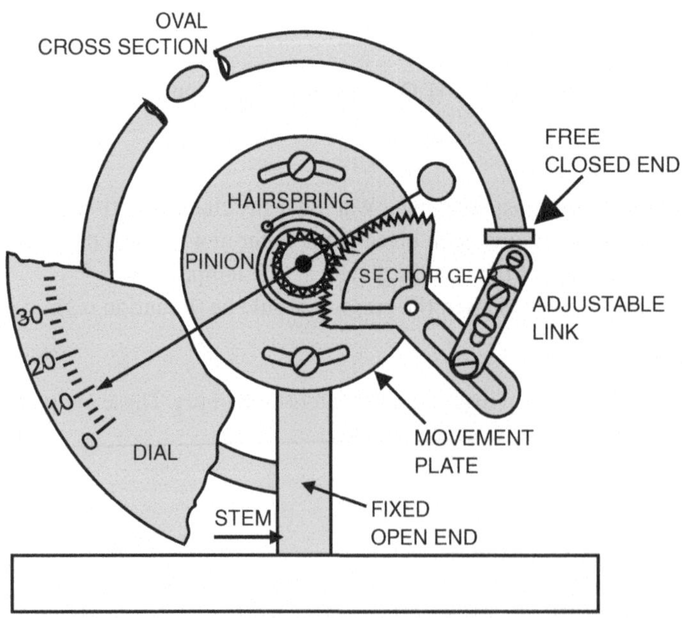

Figure 6.5 Bourdon pressure gauge.

Figure 6.6 Pressure gauge. Jim Barber. Adobe Stock Photos.

into the pointer's movement through links, levers, and gears (Figure 6.5). The measurement values of the gauge are read directly on the gauge scale from the position of the pointer (Figure 6.6).

Pressure gauges tend to have some hysteresis errors due to friction and elastic elements (Figure 6.7).

6.3.2 Pressure Transmitters

The purpose of the pressure transmitter is to sense pressure and convert it to a signal that an indicator, recorder, or controller can use. Pneumatic transmitters convert the measured pressure to a pneumatic signal, such as 3–15 psig. Electronic transmitters might have a digital output, such as a fieldbus, or an analog output, such as 4–20 mA or 1–5 V.

Unlike electronic temperature transmitters that are connected to a separate sensor, most electronic pressure transmitters include the sensor. A diaphragm isolates the sensor from the process being measured, and it flexes with changes in pressure. The transmitter's electronics convert the sensor signal into a signal that the control system can use. There are many types of pressure sensors:

- **Capacitive** – Capacitance changes as the distance changes between the diaphragm and the capacitor plates.
- **Strain gauge** – The strain gauge is affixed to the diaphragm; the resistance changes as the strain gauge flexes.
- **Inductive** – Diaphragm movement causes a metallic core to move within magnetically coupled coils, changing the inductance of the coils.

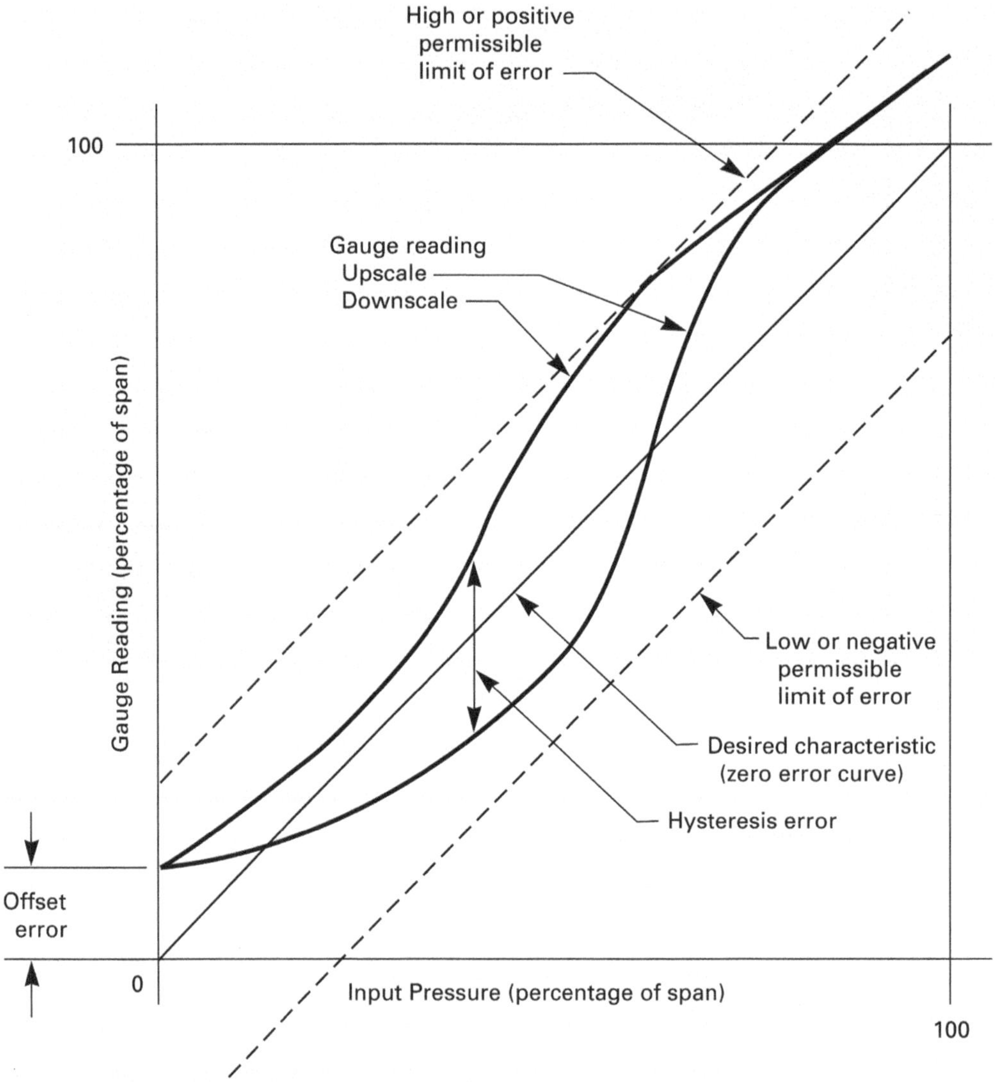

Figure 6.7 Hysteresis and offset error [1]. Reprinted from ASME B40.100-2022 by permission of The American Society of Mechanical Engineers.

For calibration purposes, it is not usually necessary to know what type of sensor is incorporated into a transmitter. The transmitter electronics handle the conversion of capacitance, inductance, or resistance to a standardized output signal.

6.3.3 Smart Transmitters

Smart transmitters incorporate three subsystems:

1) The input subsystem converts the sensor measurement signal into a digital signal (called analog-to-digital conversion or A/D).
2) The signal conditioning subsystem accepts this digital signal and performs various conditioning and mathematical manipulations to produce a digital representation of the measurement.
3) The output subsystem converts the digital signal to a robust analog output signal (D/A).

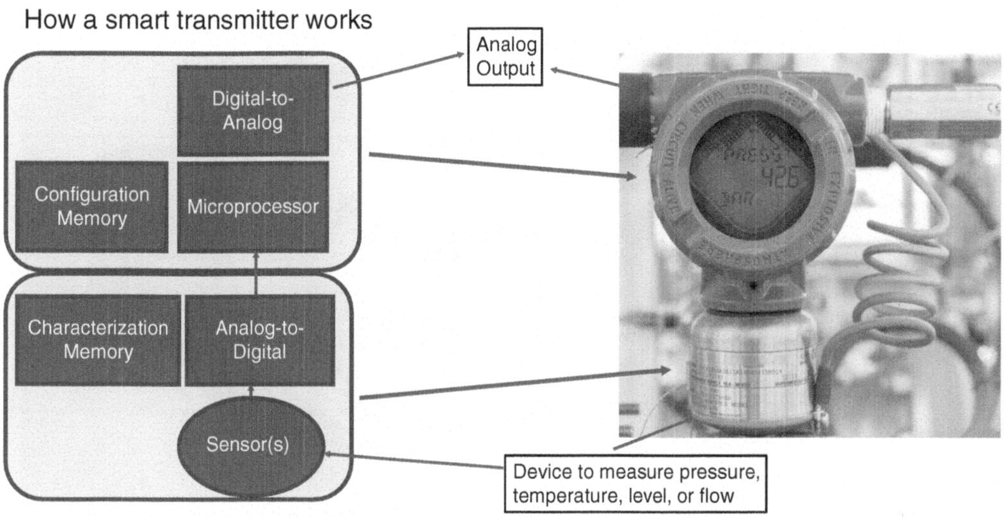

Figure 6.8 Smart transmitter. Croc80. Adobe Stock Photos.

Many manufacturers separate these into two sections, as shown in Figure 6.8:

1) Sensor section
   - Process variable sensor and other sensors
   - Characterization in a memory chip; sensors are individually characterized at the factory
2) Transmitter
   - Microprocessor
   - Configuration memory

The manufacturer's specifications will list the uncertainty associated with the A/D and D/A conversion. This should be included in the calculation for the total probable error (TPE).

6.3.4 Pressure Switches

A pressure switch is a device that senses pressure and changes states at the programmed or adjusted set point. A single-pole, double-throw (SPDT) pressure switch (Figure 6.9) has one set of normally open (NO) and one set of normally closed (NC) contacts. A double-pole, double-throw (DPDT) pressure switch is furnished with two SPDT switches. NO and NC refer to the shelf state, which is the state of the contact with the switch depressurized.

Pressure switches can be electronic or mechanical. Switches are incorporated into some gauges (Figure 6.10).

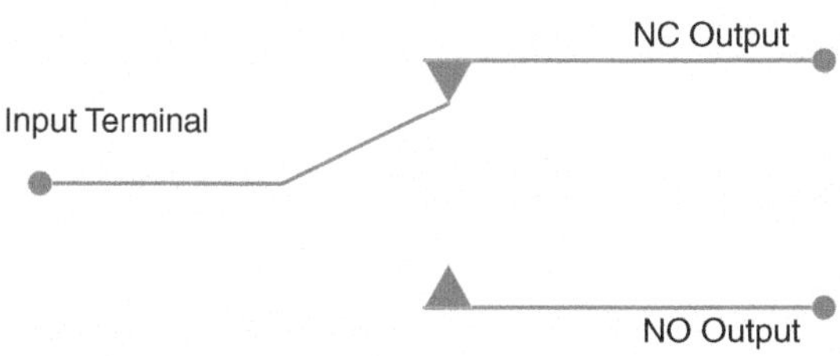

Figure 6.9 Single-pole, double-throw (SPDT) switch. NC, normally closed; NO, normally open.

Figure 6.10 Pressure gauge with switch contacts. rostovdriver. Adobe Stock Photos.

6.4 Pressure Measurement Uncertainty

CCST Level III

6.4.1 Introduction

The uncertainty in a pressure loop is very much like the discussion of temperature loop uncertainty in Section 5.4. For the experienced technician organizing a calibration program, this is important when assessing loop tolerance. Technicians performing calibration and troubleshooting should also understand the contribution that each device can make to the overall error.

Figure 6.11 is an example of a pressure measurement loop using a transmitter connected to a programmable logic controller (PLC) input card. This example is used throughout this chapter.

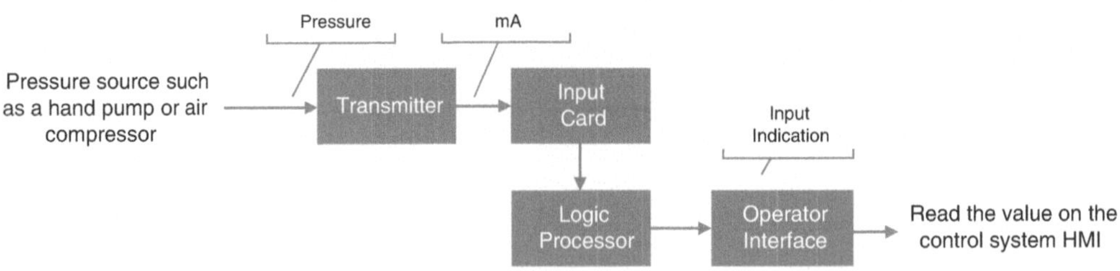

Figure 6.11 Pressure measurement loop using a transmitter connected to a PLC input card. PLC, programmable logic controller.

There will be uncertainty in both the transmitter and the input card. The TPE of the reading seen on the operator interface can be calculated using the root sum squared (RSS) equation for the combination of device errors (Equation 6.4).

$$\text{indication error} = \sqrt{(\text{transmitter error})^2 + (\text{input error})^2} \tag{6.4}$$

The uncertainties associated with each device can be found in the manufacturer's literature. It is usually necessary to use the RSS equation to combine multiple sources of error for each device to determine the likely error associated with each one.

Here is an example of calculating the uncertainty for the loop in Figure 6.11. The conditions for this example are as follows:

- Rosemount 2088 absolute and gauge pressure transmitter, range 3 (see the specifications in Appendix C)
- 4–20 mA transmitter output to the PLC analog input card
- Transmitter calibrated span: 0–200 psig = 4–20 mA
- Ambient temperature: 0 °F to 100 °F
- Maximum expected process temperature: 150 °F
- Last calibration for all devices: less than 5 years
- Transmitter upper range limit (URL): 800 psig
- Input card accuracy: 0.1% span
- Specified loop tolerance ±1.0% of span = ±2.0 psig

6.4.2 Transmitter

Multiple error components can contribute to the overall accuracy of transmitters. These include the accuracy of the electronics, the resolution of the D/A conversion, and the effect of ambient temperature on the electronics.

Table 6.2 is an example of calculating TPE for a pressure transmitter (see the Rosemount 2088 data sheet in Appendix C, Figure C.5). The manufacturer's published numbers are valid within a stated period of stability. To account for drift over time, manufacturers will sometimes give a stability number.

Calculate the TPE for the transmitter (Equation 6.5).

$$\text{TPE} = \sqrt{0.15^2 + 1.5^2 + 0.8^2} = \pm 1.7\ \text{psig} \tag{6.5}$$

Table 6.2 Pressure transmitter TPE calculation.

	Specifications	**Calculated Value for This Example**
Reference accuracy	±0.075% of calibrated span Includes the combined effects of linearity, hysteresis, and repeatability	±0.15 psig
Ambient temperature effect (t = ambient temperature)	±(0.15 percent URL + 0.15 percent span) Total effect per 50 °F	±1.5 psig
Stability	±0.1% of URL for 3 years	±0.8 psig

TPE, total probable error; URL, upper range limit.

6.4.3 Calculating Loop TPE

Refer to the discussions in Section 5.4.3, PLC Input Card, and Section 5.4.4, Calculating Loop TPE/Loop Tolerance.

For the PLC input card, use ±0.1% of the span (0.1% of 200 psig = 16 mA), which is typical for these devices. This would be ±0.2 psig.

Use the RSS equation to calculate the TPE for this example pressure loop (Equation 6.6).

$$\begin{aligned} \text{TPE} &= \pm\sqrt{(\text{transmitter})^2 + (\text{input card})^2} \\ &= \pm\sqrt{1.7^2 + 0.2^2} \\ &= \pm 1.71 \text{ psig} \end{aligned} \tag{6.6}$$

This is within the specified loop tolerance of ±2.0 psig.

6.5 Pressure Calibration

6.5.1 Introduction

Figure 6.12 shows an example of a pressure calibration setup for a pressure transmitter. If a loop calibration is performed, the output standard and power supply might not be needed. To conduct a pressure calibration, a pressure source is connected to the device under test (switch, gauge, or transmitter) in parallel with a calibrated pressure reference standard. The pressure source and reference standard might be built into a multifunction calibrator. The output is measured and recorded with a calibrated standard.

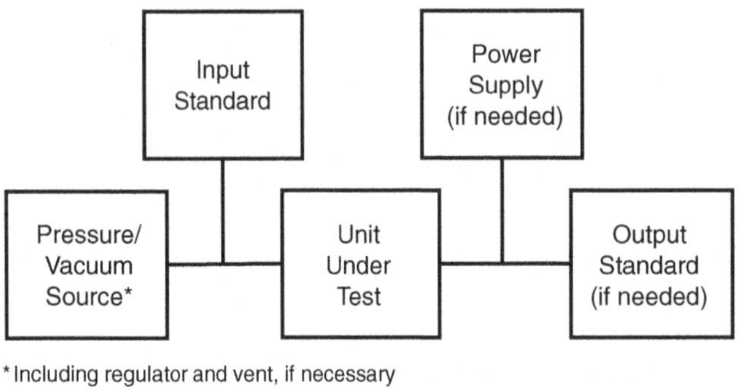

Figure 6.12 Typical pressure device calibration setup.

6.5.2 Pressure Calibration Equipment

Calibrating pressure instruments requires a pressure source to simulate and test various pressure levels on the bench or in the field.

Deadweight testers (Figure 6.13) are accurate and reliable for bench calibration. The device operates by placing calibrated weights onto a piston-cylinder assembly; the weight applies downward force, and the piston area translates this force into pressure.

Figure 6.13 Deadweight tester [2]. Reproduced with permission of Fluke Corporation.

Tech Tip

How a Deadweight Tester "Measures" Pressure

A deadweight tester does not actually measure pressure. Rather, it allows the operator to know or calculate the pressure when the system is in a specific state. This state, known as *flotation*, is achieved when there is enough force from the hydraulic pressure medium to support the piston in a floating state. The resulting pressure is proportional to the force applied by the mass load. The mass produces a force because it is accelerated by gravity—a phenomenon that varies with, and is resident to, geographical location.

When the piston is floating, the pressure (P) is derived by calculating the force (F) derived from the mass, equal to the sum of the piston mass and operator-loaded mass platters, divided by the area (A) of the piston: $P = F/A$. Because the pressure is proportional to the force, the operator measures higher pressure by loading additional mass and then reestablishing the floating equilibrium.

Source: [3].

Deadweight testers should be used in controlled conditions, so they are generally not practical for calibration in the field. Several conditions affect the accuracy of a deadweight tester. Modern testers usually have electronics or software to help compensate for these factors.

- **Gravity variation** – Deadweight testers rely on the force of gravity to generate pressure from the weights. Variations in local gravitational acceleration (e.g., due to altitude or latitude) can cause errors in pressure generation. The tester should be calibrated for the local gravity value for highly accurate calibrations.
- **Ambient temperature** – Temperature changes can affect the density of the medium (e.g., oil, water, or gas) used in the tester, as well as the expansion or contraction of the piston-cylinder assembly. These changes may introduce small errors in the pressure generated.

- **Air density (altitude)** – For pneumatic deadweight testers, changes in air density due to altitude or temperature can impact the generated pressure. This is particularly relevant when using gas mediums.
- **Vibration and leveling** – The tester must be placed on a stable, level surface to ensure accurate readings. Vibration or tilting can affect the weight distribution on the piston, leading to errors in pressure measurement.
- **Barometric pressure** – Changes in atmospheric pressure can influence measurements in pneumatic systems. Corrections may be needed to account for the local barometric pressure, especially in open systems.

Tech Tip

Gravity Variation

The acceleration of gravity is different at different locations on the Earth. The gravity at a location can be found in national and international databases given latitude, longitude, and elevation.

For example:

- Rio De Janeiro, Brazil: $g = 32.115$ ft./s^2
- Stavanger, Norway: $g = 32.214$ ft./s^2

So, a mass of 10 lbm would weigh 9.982 lbf in Rio but 10.013 lbf in Stavanger … a difference of 0.31%.

This can be significant if a deadweight tester is used to calibrate a field pressure calibration standard that is used to calibrate instruments in the plant. The error is passed on down the chain.

$$F = ma$$

weight = mass × acceleration of gravity

$$g_c = 32.1740 \frac{\text{lbm}\ \text{ft.}/\text{s}^2}{\text{lbf}} \text{ (average)}$$

Field calibration is done with a portable pressure source. Pressure sources can be hydraulic (liquid) or pneumatic (gas/air). The pressure can be supplied by a hand pump, a gas cylinder, or a mechanical source. Each type of pressure source has its strengths and limitations, and the choice depends on the specific pressure range, environment, portability, and frequency of use required for the calibration task (Table 6.3).

Hand pumps (Figure 6.14) are manually operated and can be either hydraulic or pneumatic. They are portable and lightweight and don't require a power source. Hand pumps often come with fine-adjustment controls for precise pressure setting, and they are typically limited to lower-pressure ranges than mechanical pumps.

Mechanical pumps (Figure 6.15), such as electric or pneumatic-driven pumps, provide automated pressure generation. They are often used in field applications requiring frequent calibration

Table 6.3 Hydraulic versus pneumatic pressure sources.

Hydraulic Sources	Pneumatic Sources
Use fluids, typically oils or water, to generate pressure	Use gases, usually air or nitrogen, to generate pressure
Can be used for generating high-pressure levels, often up to 10,000 psi or more	More suitable for low- to medium-pressure ranges, typically up to around 600 psi
Stable pressure, minimal air compressibility	Potential for gas compressibility effects and leaks that can make achieving stable pressure more challenging
Heavier and less portable	Lighter and more portable

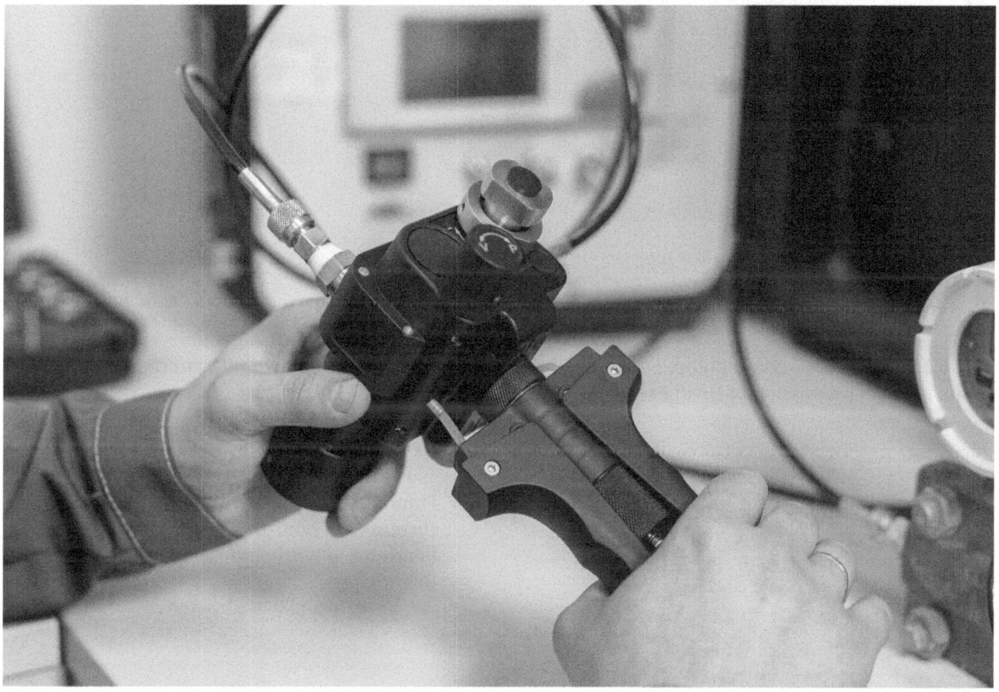

Figure 6.14 Hand pump pressure source. NEZNAEV. Adobe Stock Photos.

where manual pumping would be laborious and impractical. These require a power source and usually cost more. Portable units for lower-pressure applications can be battery powered.

Gas cylinders, used with a precision pressure regulator, provide a stable, high-pressure source of compressed gas, usually nitrogen or other inert gases. They're commonly used in pneumatic calibration for high-accuracy instruments and applications requiring stable pressure over a prolonged period. Although gas cylinders provide steady pressure, some are bulkier and less portable, requiring regulators and safety precautions. Self-contained units (Figure 6.16) with internal cylinders are more portable and can be filled in the shop to support several field calibrations.

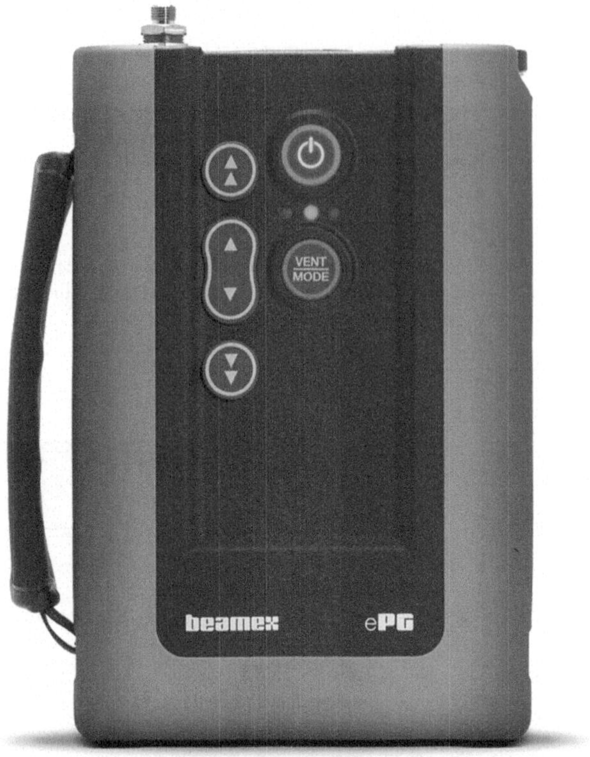

Figure 6.15 Electric pressure pump. *Beamex ePG: Electric Pressure Pump and Controller brochure* [4]. Reproduced with permission of Beamex.

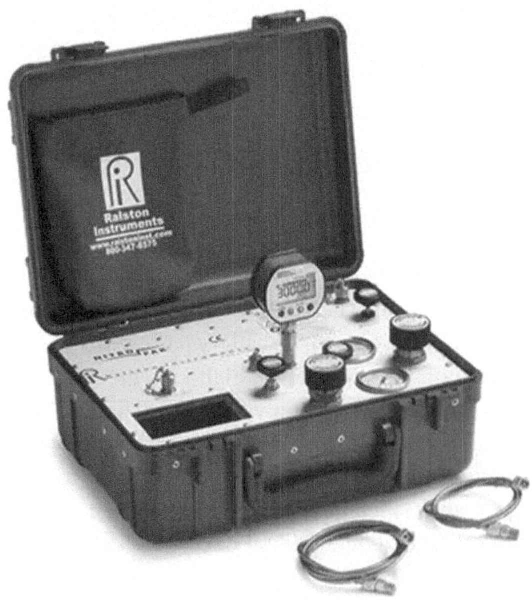

Figure 6.16 Compressed gas calibration system with built-in gas cylinder [5]. Reproduced with permission of Ralston Instruments.

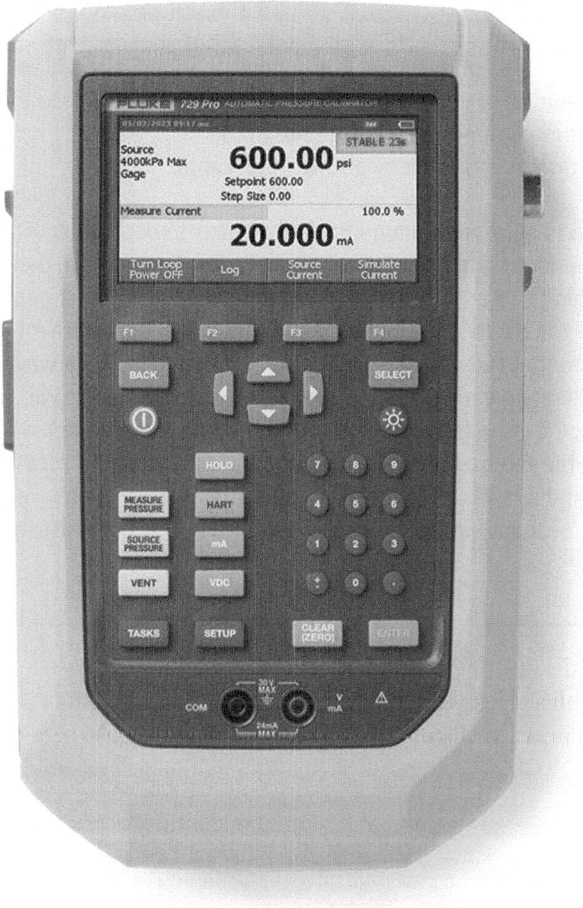

Figure 6.17 Automatic multifunction pressure calibrator [6]. Fluke 729 Pro Automatic Pressure Calibrator technical data sheet. Reproduced with permission of Fluke Corporation.

Full-function (low-pressure) automated devices (Figure 6.17) will source and measure pressure and read the transmitter output in milliamps or voltage. These have a built-in mechanical pump and can be set to run a fully automated calibration on multiple pressure test set points. These are usually limited to lower pressures.

The following is a summary of pressure source use cases:

- **Low- to medium-pressure calibration (up to ~600 psi)** – Pneumatic hand pumps are ideal for clean and portable solutions, particularly in electronic and sensitive environments.
- **High-pressure calibration (>600 psi)** – Hydraulic hand pumps, mechanical hydraulic pumps, and electric pumps are preferred for high-pressure applications, like calibrating high-range gauges and transmitters.
- **Frequent or high-precision calibration** – Mechanical pumps (either pneumatic or hydraulic) provide consistent pressure generation with less manual effort, making them suitable for both field and lab settings where repetitive, high-precision work is needed. Gas cylinders can also support frequent field calibrations.

Tech Tip

Pressure calibrators typically support many units of measure. It is important to understand the relative pressure units (gauge versus absolute) to ensure the correct test equipment, pressure module, and tubing/fittings rated for the maximum test pressure are used.

Here is an anecdote from the first edition of this book by author Mike Cable:

> When I first started calibrating, I had a pressure gauge that read about 15 psi high at all test points. I went through a tremendous amount of effort to drain the glycerin from the gauge, remove the cover, and attempt to adjust the gauge needle. When I couldn't get the gauge to calibrate properly, I took it to my supervisor to order a new replacement gauge. My supervisor informed me that the gauge read in psia, my "as-found" data was acceptable, and I ruined a perfectly good gauge. The gauge was ruined because the vacuum that had existed in the gauge case was eliminated when the gauge was opened. This vacuum was present to provide the gauge with a reference to absolute zero pressure. So, if you calibrate a gauge and it reads about 15 psi with no pressure applied, take it from me, the gauge reading is probably in psia, not psig. Know the unit of measure for the device you are calibrating!

Pressure devices are usually installed with a block valve and a bleed valve or a manifold that can block system pressure and relieve trapped pressure (Figure 6.18). The bleed valve can also be used to connect a pressure calibrator.

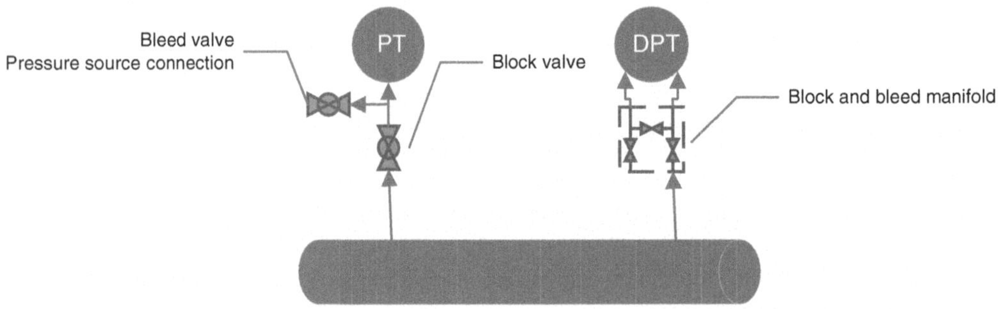

Figure 6.18 Pressure device installation with block and bleed. DPT, differential pressure transmitter; PT, pressure transmitter.

Tech Tip

From the first edition of this book, author Mike Cable:

Another challenge with pressure calibration is having the correct fittings to connect the pressure source, pressure standard, and the unit under test. To minimize this problem, include a list of the fittings needed in the calibration procedure. It is important to minimize the number of connections in the test setup to minimize the potential for leakage.

6.5.3 Safety

Safety considerations should be observed when calibrating pressure instruments:

- Ensure the device is isolated from the system pressure and any trapped pressure has been released before attempting to remove any pressure devices.
- Know the potential hazards that exist with the process material. Some process materials may be toxic or cause harm to the environment if released.
- Communicate with the area operations supervisor before removing any device for calibration.

At lower pressures, pressure devices can be checked in place using air or other gas and a portable calibrator. Using a gas at high pressure can be dangerous because compressed gas stores a lot of energy. If a connection or tubing fails, the energy will be released suddenly and can cause harm. Hydraulic testing is safer for higher-pressure calibration.

6.5.4 Pressure Calibration Examples

6.5.4.1 Pressure Calibration: Pressure Gauges

An analog pressure gauge might require periodic calibration checks because of its links, levers, and elastic pressure sensing element. Calibrating a pressure gauge can include adjusting these components until the gauge accurately represents the input.

In this test, we will calibrate this pressure gauge (Figure 6.19): Ashcroft 1279, 0–200 psig, ASME Grade 2A (accuracy ±0.5% full scale permissible error).

The reference standard is a Fluke 700GA27 pressure test gauge, 0 psig to 300 psig model, with an overall uncertainty of 0.05% of full scale range, attached to the test pump (Figure 6.20).

Assume there is no specified tolerance for this gauge, and test it against the manufacturer's accuracy specification (±0.5% full scale).

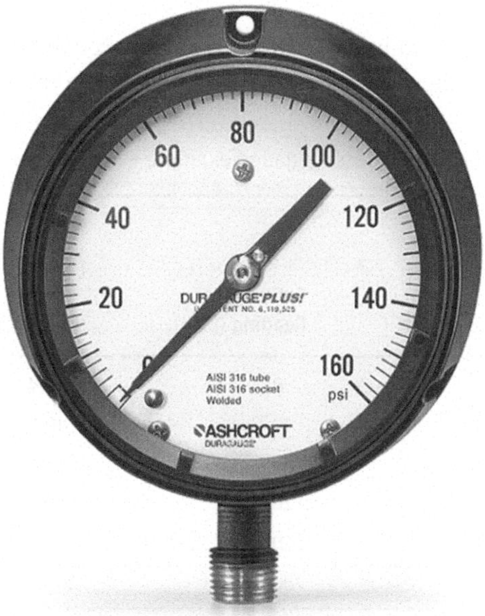

Figure 6.19 Ashcroft 1279 pressure gauge [7]. Reproduced with permission of Ashcroft Inc.

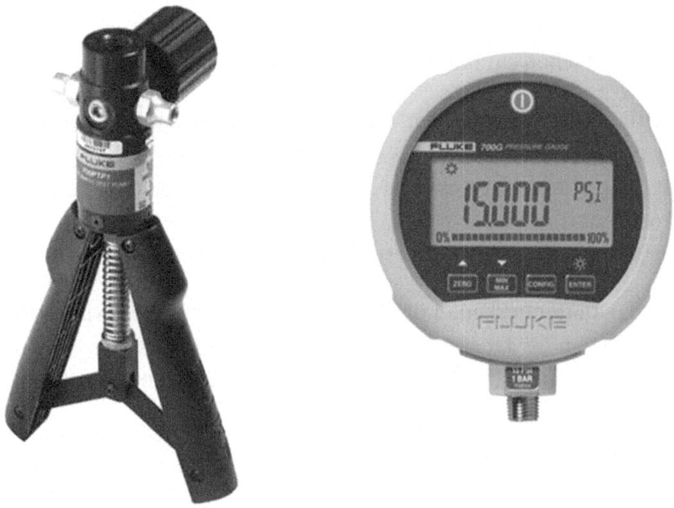

Figure 6.20 Fluke 700PTP-1 pneumatic test pump [8] and 700G test gauge [9]. Fluke 700PTP-1 Pneumatic Test Pump and 700G Series Precision Pressure Test Gauges technical data sheets. Reproduced with permission of Fluke Corporation.

Calculate the test accuracy ratio (TAR) as called for in the calibration procedure (Equation 6.7):

$$\text{TAR} = \frac{\text{tolerance of device}}{\text{accuracy of reference standard}}$$

$$\text{gauge tolerance} = 200\ \text{psig} \times 0.5\% = 1\ \text{psig}$$

$$\text{calibrator accuracy} = 300\ \text{psig} \times 0.05\% = 0.15\ \text{psig}$$

$$\text{TUR} = \frac{1\ \text{psig}}{0.15\ \text{psig}} = 6.7:1 \tag{6.7}$$

The calibrator TAR >4:1, so it is acceptable for calibrating this gauge.

Follow the procedure for calibrating a pressure gauge in Appendix D.

Table 6.4 Pressure gauge example – gauge readings [10].

		Gauge Readings (psig)			
		Increasing		Decreasing	
Test Point	Calibrator Reading (psig)	Before Tap	After Tap	Before Tap	After Tap
0%	0	−0.4	0.0	−0.4	0.0
20%	40	40.8	41.0	41.4	41.1
40%	80	80.4	80.5	81.2	81.0
60%	120	119.6	119.0	120.8	120.6
80%	160	159.2	159.5	160.6	160.4
100%	200	200.4	200.8	200.4	200.4

Example data used with permission from Ashcroft.

Be sure the gauge under test is mounted in the same orientation as it would be in the process. Positioning a gauge vertically for calibration and then reinstalling it in a horizontal configuration in the process will introduce an error.

Note that the procedure calls for taking a reading, giving the gauge a light tap, and then taking the reading again to overcome friction.

The test results (Table 6.4, Table 6.5, and Figure 6.21) are then checked against the allowable tolerance. If the results are outside the allowable tolerance, determine the type of errors present (i.e., linearity, zero, span, and hysteresis), and make adjustments according to the manufacturer's instructions. In this case, there are points out of tolerance, and the error appears to be nonlinear. The Ashcroft 1279 has no adjustments, so the gauge would be considered for replacement. If a gauge requires adjustment, follow the manufacturer's instructions.

Table 6.5 Pressure gauge example – gauge errors.

		Gauge Error (%FS)			
		Increasing		Decreasing	
Test Point	Calibrator Reading (psig)	Before Tap	After Tap	Before Tap	After Tap
0%	0	−0.20%	0.00%	−0.20%	0.00%
20%	40	0.40%	0.50%	0.70%	0.55%
40%	80	0.20%	0.25%	0.60%	0.50%
60%	120	−0.20%	−0.50%	0.40%	0.30%
80%	160	−0.40%	−0.25%	0.30%	0.20%
100%	200	0.20%	0.40%	0.20%	0.20%

%FS, percentage of full scale.
Example data used with permission from Ashcroft.

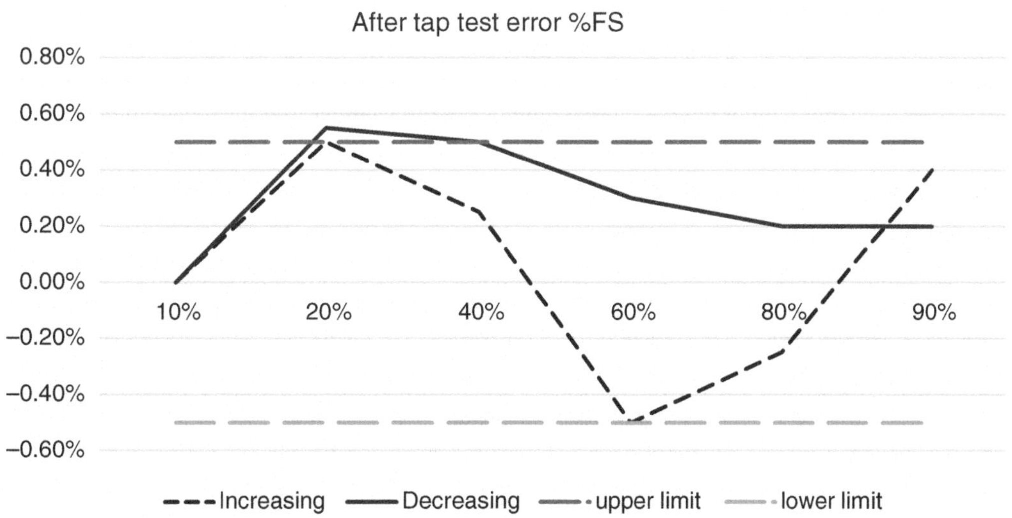

Figure 6.21 Pressure gauge calibration example results graph. %FS, percentage full scale. Example data used with permission from Ashcroft.

6.5.4.2 Example Pressure Calibration: Pressure Loop

Refer to the example pressure loop in Figure 6.11 and the pressure loop procedure in Appendix C. This example will use the uncertainty calculation for the loop in Section 6.4.

For this example:

- Pressure transmitter: Rosemount 2088 absolute and gauge pressure transmitter (range 3 – see the specifications in Appendix C).
- Transmitter calibrated span: 0–200 psig = 4–20 mA.
- Ambient temperature during calibration: 0 °F to 100 °F.
- PLC input card: 4–20 mA input (sourcing) with stated accuracy ±0.1% of span.
- Loop TPE: ±1.71 psig.
- Required loop tolerance = ±2.0 psig.
- Calibration equipment: Fluke 729 (specifications in Appendix E). This device has an internal electric pump and an internal reference standard for measurement. It will automatically run the full test after the operator enters the test points.

Verify the TUR. The TPE for the calibrator reference standard must be calculated at the calibration conditions. The test is performed outdoors, and the ambient temperature can range from 0 °F to 100 °F.

Calibrator overall uncertainty: The 300 psi module is 0.02% full scale/year, maximum in the operating range of 14 °F to 122 °F (−10 °C to +50 °C) per data sheet. Because the actual conditions fall within the specified operating range, no correction to the uncertainty is needed.

Loop tolerance is specified as a fixed value (±2 psig) across the entire range, so the upper specification limit (USL) and lower specification limit (LSL) must only be calculated at one point.

$$\text{USL} = 200\ \text{psig} + 2\ \text{psig} = 202\ \text{psig}$$

$$\text{LSL} = 200\ \text{psig} - 2\ \text{psig} = 198\ \text{psig}$$

$$u = 300\ \text{psig} \times 0.02\% = 0.06\ \text{psig}$$

$$\text{TUR} = \frac{\text{USL} - \text{LSL}}{2u} = \frac{202 - 198}{2 \times 0.06} = 33$$

Because 33 is more than 4, this TUR is acceptable.

Follow the steps in the calibration procedure and record the as-found data points, reading the loop values from the operator interface. (For simplicity in this example, the applied input, measured by the 729 internal reference, is exactly the specified value.)

The results are in Table 6.6. All readings have an error within the specified tolerance of ±2 psig.

Table 6.6 Pressure loop calibration results.

			As-Found		
Input (%)	**Applied Input (Standard)**	**Desired Indication**	**Increasing Indication**	**Error in Engineering Units**	**Error in Reading (%)**
0	0 psig	0 psig	0 psig	0 psig	0
25	50 psig	50 psig	50.5 psig	+0.5 psig	1.0
50	100 psig	100 psig	100.8 psig	+0.8 psig	0.8
75	150 psig	150 psig	100.7 psig	+0.7 psig	0.47
100	200 psig	200 psig	201 psig	+1 psig	0.5

Tech Tip

It might be advantageous to record the current output of the transmitter in addition to the human-machine interface (HMI) reading. If the loop is out of tolerance, this would save the step of running calibration on the pressure transmitter as a troubleshooting step.

6.5.4.3 Example Pressure Calibration: 4–20 mA Pressure Transmitter

Calibrating a pressure transmitter is like calibrating a pressure gauge, except the transmitter's output signal must be measured using an appropriate measuring standard. The following equipment is needed for the device calibration:

- Pressure source
- Pressure reference standard
- Power supply
- Ammeter to read the loop current

This example will use the same Fluke 729 used in the example in Section 6.5.4.2 and connected as shown in Figure 6.22. The Fluke 729 performs all the necessary functions, so four separate pieces of equipment are not needed.

Adjust the input values to the required test points, and record as-found values for the milliamp output. If the transmitter includes a local readout, record the value on the transmitter's display. Typically, it is unnecessary to check a transmitter in both the upscale and downscale directions, and this procedure does not call for it. Transmitters are less likely to exhibit hysteresis than gauges are.

Once the as-found data has been recorded (Table 6.7), evaluate the error against the acceptable tolerance to determine if adjustments are required. If necessary, perform the zero and span

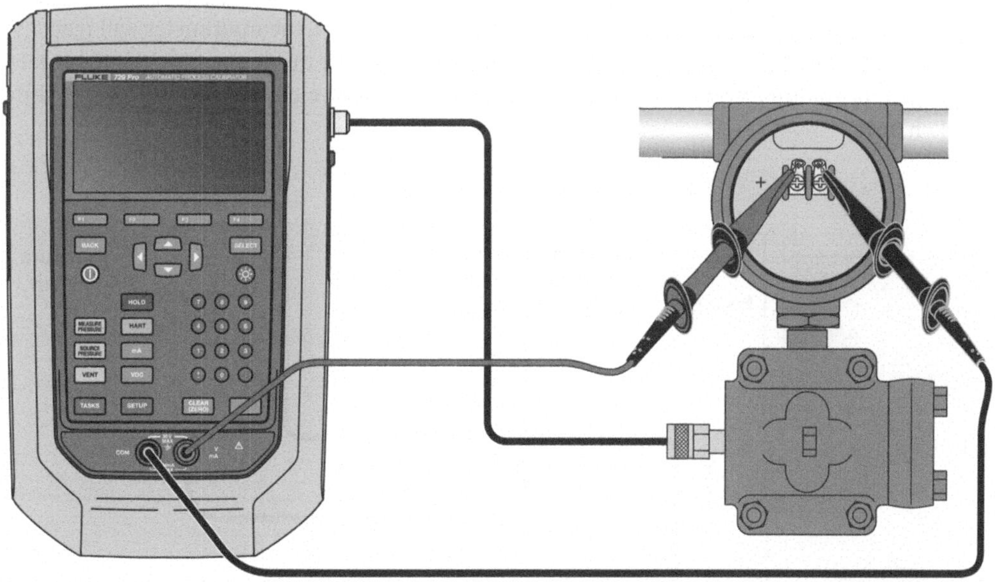

Figure 6.22 Pressure transmitter calibration hookup. *Fluke 729 PRO Automatic Pressure Calibrator Users Manual*, Rev 2, Figure 5 [11]. Reproduced with permission of Fluke Corporation.

Table 6.7 As-found data.

			As-Found		
Input (%)	**Applied Input (Standard)**	**Desired Indication**	**Increasing Indication**	**Error in Engineering Units**	**Error in Reading (%)**
0	0 psig	4.0 mA	4.000 mA	0	0
25	50 psig	8.0 mA	8.080 mA	+0.08	1.0
50	100 psig	12.0 mA	12.096 mA	+0.096	0.8
75	150 psig	16.0 mA	16.075 mA	+0.7	0.47
100	200 psig	20.0 mA	20.100 mA	+0.1	0.5

adjustments according to the manufacturer's procedures. This may involve adjusting the zero and span potentiometer, pressing a button, or using a handheld interface. Because some zero and span adjustments interact, it is important to recheck the zero and span until no further adjustment is required.

Once no further adjustments are required, repeat the 5-point check and record the as-left values.

6.5.4.4 Example Pressure Calibration: Pressure Switch

The calibration setup for a pressure switch is similar to that of a pressure gauge and pressure transmitter except that a multimeter set to measure voltage or resistance, as applicable, is connected to the appropriate set of contacts (such as the common and NO terminals), as shown in Figure 6.23. In this example, the meter is incorporated into the Fluke 729 multifunction calibrator.

The applied input pressure is increased to the set point. The multimeter will read the change of voltage or resistance when the switch changes state. In other words, the multimeter will read 0 V with the switch closed and supply voltage with the switch open. If the contact is a "dry" contact (no voltage), the multimeter will read close to 0 Ω with the contact closed and infinite ohms with the switch open.

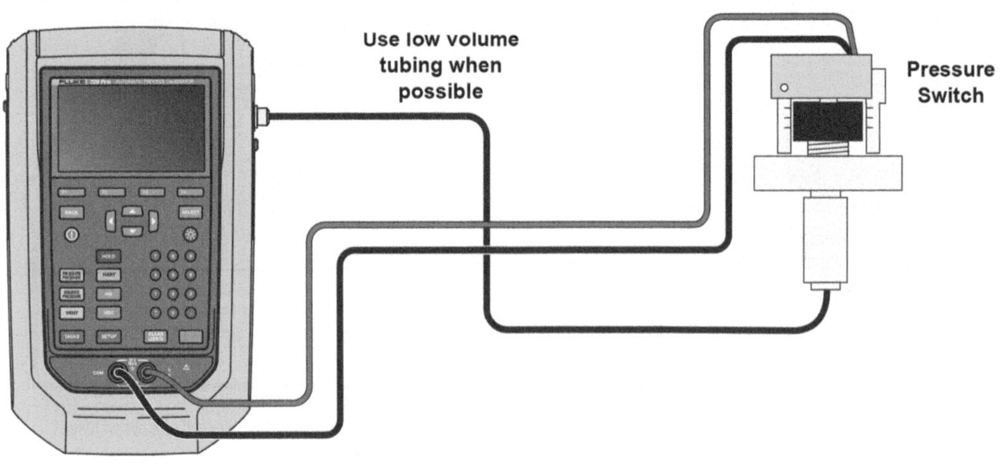

Figure 6.23 Pressure switch calibration example. *Fluke 729 PRO Automatic Pressure Calibrator Users Manual*, Rev 2, Figure 6 [12]. Reproduced with permission of Fluke Corporation.

Tech Tip

The tricky part of calibrating a pressure switch is to know whether the switch should trip with increasing or decreasing pressure and which contact is used for the circuit.

Typically, a high-pressure switch would trip with increasing pressure and would use the NC contact. This means that the switch would be closed at normal pressure and would open when the high-pressure set point is reached. Using the NC contact provides a fail-safe condition if the circuit fails (power failure or broken wire).

Review Questions

6.1 If a gauge pressure device reads 110 psig, what would the absolute pressure reading be if the device were in:

a) Denver, Colorado (atmospheric pressure = 12.1 psia)?

b) Houston, Texas (atmospheric pressure = 14.7 psia)?

6.2 Using the piping and instrumentation diagram (P&ID; DRG P&ID 15) in Appendix A, identify the components in pressure loop 303. List the tags. What is the type and function of each device?

6.3 There would be two calibration procedures for the loop in Question 6.2. One for the input side and one for the output side of the controller. List the components that would be in each procedure.

6.4 Does the transmitter in Question 6.2 have both a block valve and a bleed valve to allow attaching the pressure source without removing the transmitter from service?

6.5 Summarize the main steps in calibrating the input loop identified in Question 6.3.

6.6 If a loop tolerance is ±1 psig, what is the highest desirable value for the TPE of the calibrated device used to read the applied pressure?

6.7 A pressure gauge on a nozzle near the bottom of a tank reads 5.0 psig. The tank is open to the atmosphere, so there is no blanket gas pressure. The specific gravity of the fluid in the tank is 0.8. Calculate the height of the fluid in the tank.

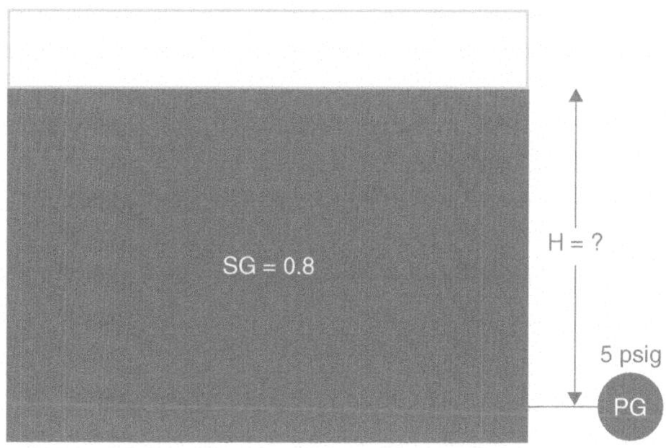

6.8 What test equipment would be needed to calibrate an electronic pressure transmitter?
   a) Pressure standard, current standard
   b) Current standard, communicator
   c) Pressure standard, analog voltmeter
   d) Pressure standard

6.9 If an elephant weighing 12,000 lb. stands on one foot that has an area of 200 in.^{2}, how much pressure is under that foot?

6.10 If a human weighing 150 lb. stands on one foot that has an area of 30 in.^{2}, how much pressure is under that foot?

6.11 If the elephant stands on the human's foot, then what?

References

Note: The dates and editions for standards and other references in this chapter are current as of publication. The reader is encouraged to check for later revisions or editions.

1 American Society of Mechanical Engineers (ASME), ASME B40.100-2022, *Pressure Gauges and Gauge Attachments*. New York, NY: ASME, 2022.

2 Fluke, 2011–2013, *Fluke P3000 Series Hydraulic Deadweight Testers Technical Data*. Fluke Corporation. https://media.fluke.com/3003d8f7-7768-420d-9142-b2e70078631d_original%20file.pdf. Accessed February 4, 2026.

3 Grossman, Jeff, "The Electronic Deadweight Tester," *Cal Lab Magazine: The International Journal of Metrology* 16.3 (2009), 36–39.

4 Beamex, 2024, December, *Beamex ePG: Electric Pressure Pump and Controller* brochure. Beamex. https://www.beamex.com/calibrators/beamex-epg. Accessed February 4, 2026.

5 Ralston, 2024, *NitroPak Pressure Test Kit*. Ralston Instruments. https://www.ralstoninst.com/nitrogen-pressure-test-kit. Accessed December 2024.

6 Fluke, 2024, *Fluke 729 Pro Automatic Pressure Calibrator Technical Data Sheet* (03/2021). Fluke Corporation. https://media.fluke.com/20ee6c2a-91cd-46f7-9ec2-b3190159abf3_original%20file.pdf. Accessed December 2024.

7 Ashcroft, 2024, *Ashcroft 1279 Duragauge Pressure Gauge* data sheet. Ashcroft Inc.

8 Fluke, 2004, *Fluke 700PTP-1 Pneumatic Test Pump Technical Data Sheet*. Everett, WA: Fluke Corporation.

9 Fluke, 2013, *Fluke 700G Series Precision Pressure Test Gauges Technical Data Sheet*. Everett, WA: Fluke Corporation.

10 Ashcroft, 2015, December, *Pressure Gauge Installation, Operation, and Maintenance*, I&M008-10098-5/02 Rev. C. Stratford, CT: Ashcroft Inc.

11 Fluke, 2020, September, *Fluke 729 PRO Automatic Pressure Calibrator Users Manual*. Everett, WA: Fluke Corporation.

12 Fluke, 2020, September, *Fluke 729 PRO Automatic Pressure Calibrator Users Manual*. Everett, WA: Fluke Corporation.

7

Level Instrument Calibration

7.1 What Is Level?

Measuring level is about finding the position of a surface inside a container. For example, what is the height of water in a tank or grain in a silo? It can also mean finding the level of an interface between two fluids, such as oil and water, in a separation vessel. See Figure 7.1.

Why is level measured? Some reasons include keeping track of inventory, measuring material quantity, preventing overflow or underflow, controlling a process, and detecting leaks.

7.2 Level Terms and Units of Measure

Frequently used units of measure for pressure are listed in Table 7.1. Level is reported in units of distance, percentage, or volume.

Level is often given as the percentage of the overall height of the storage tank, vessel, or silo instead of providing the height in units of distance. Sometimes, it is given as a percentage of the measurement span if the measurement is only for a portion of the overall vessel height. The measured span is the distance between the lowest and highest level that a level transmitter can measure in the container. See Figure 7.2.

The *reference gauge point* in a tank is a specific location used as the reference level for measurement and calibration purposes. It is the point to which the measured levels are compared or from which they are calculated. It is usually at or near the base of the tank, or it might be the lowest point that can be measured.

Innage and ullage (also called *outage*) are terms commonly used in industries such as oil and gas, shipping, and storage to describe the measurement of liquid in a tank.

- *Innage* is the measurement of the height or depth of the liquid from the bottom of the tank to the liquid's surface. It is the quantity of liquid present in a container.
- *Ullage* (or *outage*) is the measurement of the empty or free space above the liquid in a container (from the liquid's surface to the top of the tank). It is the space not filled with liquid.

Although innage is usually the measurement of interest, top-down measurement devices usually measure ullage, and then innage is inferred from the ullage measurement.

Level measurement can be used to calculate the volume of fluid in the container by using a formula or a tank capacity table (also known as a *strapping table*), which relates the level to the corresponding volume. The level measurement might be shown as a volume instead of a level on a human-machine interface (HMI) display. Accurate volume is required for custody transfer and inventory applications; accurate level measurement is necessary to calculate an accurate volume.

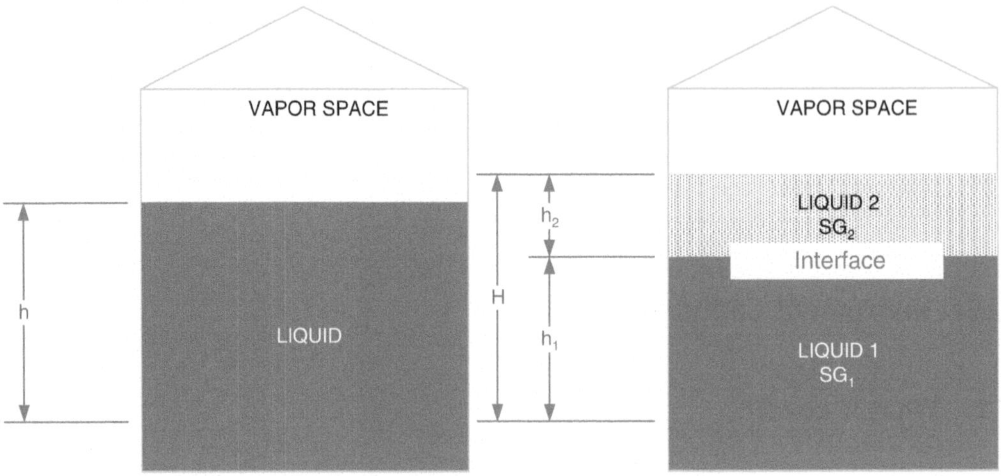

Figure 7.1 Level and interface level measurements. H, height; SG, specific gravity.

Table 7.1 Units of measure for level.

SI Standard Unit of Measure	
Meter	m
Other Common Units of Measure	
Feet	ft.
Inches	in.
Percentage full	%
Percentage of measured span	%
Volume units	

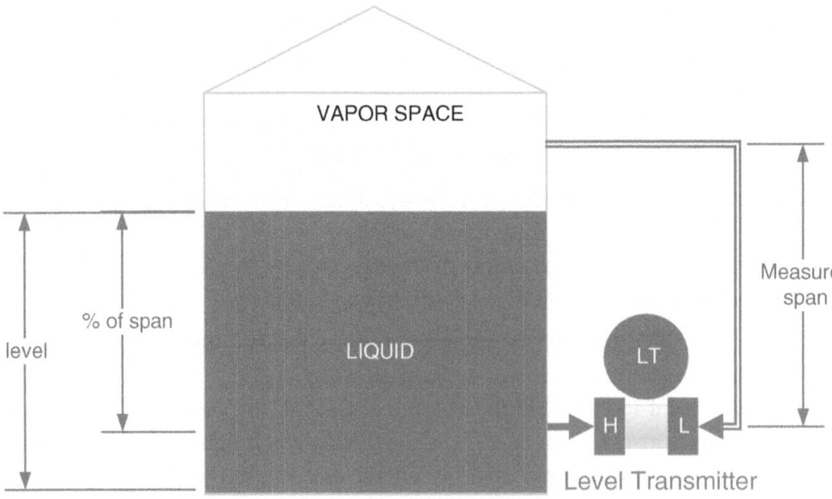

Figure 7.2 Level terms. H, high; L, low; LT, level transmitter.

7.3 Types of Level Instruments

There are numerous types of level sensors and instrumentation for both continuous analog level measurement (level transmitters) and point-level sensing (level switches). This chapter focuses on continuous-level transmitters requiring calibration. Some types of point-level measurement switches also require calibration for the application.

Level measurements can be direct or indirect.

- *Indirect* (or *inferred*) measurements use some other property, such as pressure, to determine the level.
- *Direct* measurements directly measure the height of the material surface.

Direct measurements can be top-down or bottom-up.

- **Top-down** – The sensor is at the top of the tank, and it usually measures the ullage.
- **Bottom-up** – The sensor is at the bottom of the tank, and it usually measures the innage.

Level measurement devices fall into these main categories:

- Indication only
- Mechanical/pneumatic
- Electromechanical
- Electronic contacting
- Electronic noncontacting

The sensor type used in a particular application depends on many factors, including:

- Is the material a liquid or a solid?
- Is the vessel pressurized?
- Is there an agitator in the vessel?
- Is an analog measurement needed, or just detection at a point?
- Will the device need to be accessed for maintenance?

Table 7.2 summarizes typical applications where a control system technician (CST) might find the different technologies applied. A CST should understand the fundamental principles of the level technologies to be calibrated. This section will briefly describe just a few sensor types.

7.3.1 Hydrostatic Pressure

Pressure sensors are often used in liquid-level applications. A column of fluid exerts pressure that is equal to the height of the fluid multiplied by the specific gravity of the fluid.

- *Specific gravity* is the ratio of the density of the fluid to the density of water at 60 °F.

Measuring level using hydrostatic pressure measurement is based on the principle that the pressure caused by the fluid column increases as the depth increases. The hydrostatic pressure is equal to the height of the fluid multiplied by the specific gravity of the fluid (Equation 7.1).

$$\text{pressure (in. } H_2O) = \text{height (in.)} \times \text{specific gravity} \tag{7.1}$$

It is important to understand that the pressure is affected by the depth, not the volume of the liquid. The pressure from 1 ft. of water depth will be the same for a small volume or a large volume (Figure 7.3).

Table 7.2 Applications for continuous-level measurement devices.

Type of Level Transmitter	Material	Category	Typical Applications
Hydrostatic Pressure	Liquid	Indirect	• Atmospheric tanks • Closed tanks with gas blankets • Pressurized tanks or vessels • Applications without liquid density variations • Bottom-up measurement • Infer level from hydrostatic pressure
Capacitance	Liquid Solid	Indirect	• Bulk solids (powders, grains) • Liquids with consistent dielectric properties • Nonconductive liquids (e.g., oils) • Food and beverage tanks (sanitary applications with proper coatings) • Interface level
Ultrasonic	Liquid Solid	Direct	• Open and closed tanks or vessels with no gas blanket or dusty vapor space • Open channels • Applications with moderate turbulence and no heavy foam or vapor interference
Radar (Guided Wave)	Liquid Solid	Direct	• High-pressure or high-temperature tanks • Viscous or sticky materials (e.g., crude oil and slurries) • Applications with foam, vapor, or turbulence • Tall silos or long-range liquid-level monitoring • To measure the overall level plus the interface in separator vessels
Radar (Noncontact)	Liquid Solid	Direct	• Corrosive or hazardous chemicals (acid tanks) • Tall storage tanks (liquids or bulk solids) • Environments with dust or vapor • Nonintrusive installations for clean applications
Displacer or Float	Liquid	Direct	• Simple liquid-level monitoring • Clean liquids (e.g., water and oil) • Displacers can be used for measuring the interface level • Limited to steady-state environments without significant turbulence or foam
Nuclear	Liquid Solid	Indirect	• Extremely harsh conditions (radiation, high temperatures, high pressure) • Vessels with no penetrations for inserting other types of level devices • Materials with challenging properties (e.g., slurries, foam, or emulsions)
Laser	Liquid Solid	Direct	• Tall silos or tanks (e.g., grain, cement, and coal) • Noncontact applications for aggressive or corrosive materials
Magneto-strictive	Liquid Solid	Direct	• Custody transfer of liquids (e.g., oil and refined products) • LPG/LNG tanks • High-precision applications • Sanitary applications in the food, pharmaceutical, and beverage industries
Magnetic	Liquid	Direct	• Water and oil separators • Boiler level monitoring • Noncritical level indication • Where typically incorporated with a level indication in bridles attached to vessels
Servo	Liquid	Direct	• High-precision level measurement • Inventory control in oil and gas tanks • Custody transfer and fiscal measurement systems

LNG, liquefied natural gas; LPG, liquefied petroleum gas.

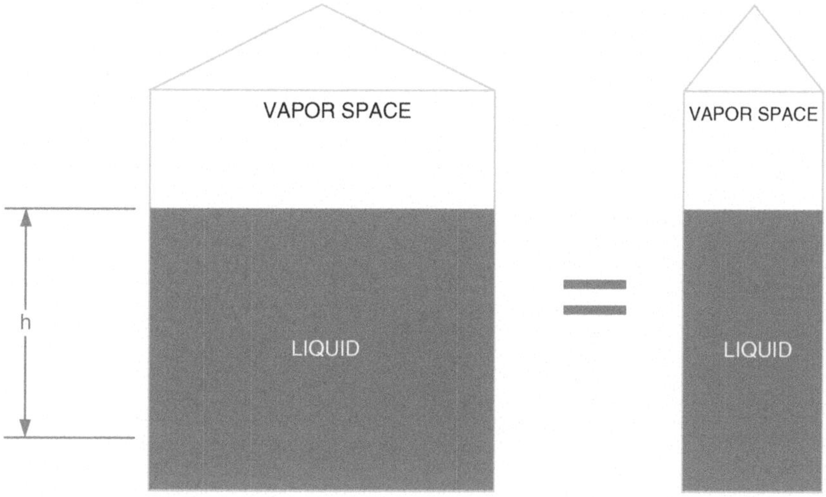

Figure 7.3 Pressure is related to depth, not volume.

Table 7.3 Level measurement by hydrostatic pressure.

Atmospheric tank	Gauge pressure transmitter (external) Submersible hydrostatic level transmitter
Pressurized tank or vessel	DP transmitter with the low side connected to the vapor space DP transmitter with remote seal and capillary tube Two pressure transmitters (one in the fluid, one in the vapor space) Transmitter with two pressure sensors

DP, differential pressure.
Note: The examples in this chapter are for a DP transmitter in a pressurized tank.

This method of level measurement is a bottom-up method. The pressure is measured at the bottom of the range. When the vessel is pressurized, the pressure in the vapor space must be subtracted from the pressure at the bottom. The pressure of the column of fluid and the vapor space pressure can be measured in multiple ways depending on whether the vessel is pressurized or vented to the atmosphere (Table 7.3).

A differential pressure (DP) transmitter has a high-pressure side and a low-pressure side, with a sensing element separating the two sides. The high-pressure side of the sensor is usually connected to the bottom portion of the vessel and, therefore, senses the pressure exerted by the process fluid in the vessel plus the tank pressure (if pressurized). The low-pressure side is usually connected to the vapor space at the top of a pressurized vessel or is open to the atmosphere if the vessel is vented or open to the atmosphere. The low-pressure side only senses the vessel pressure (or atmospheric pressure) and not the head of the process fluid (Figure 7.4).

The distance (h) is the height of the fluid (inches) above the tap where the transmitter is connected; it is not the total height of fluid from the bottom of the vessel. The pressure units are inches of water head in this example. Because the low side of the transmitter is in the vapor space, the pressure in the vapor space does not affect the level measurement (Equation 7.2).

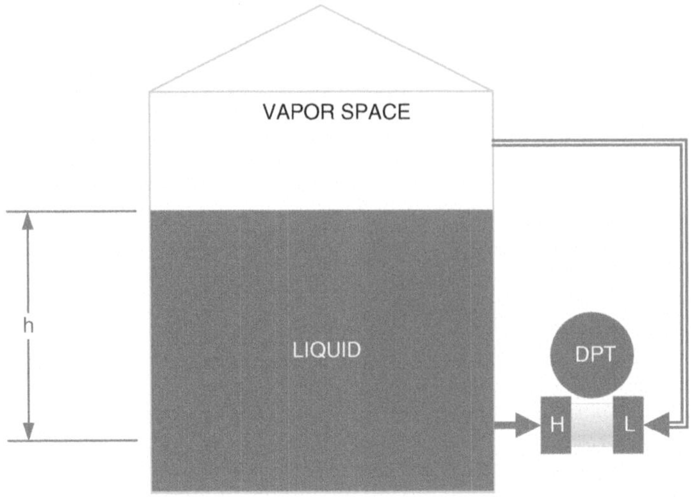

Figure 7.4 Differential pressure level measurement. DPT, differential pressure transmitter; H, high; L, low.

$$\text{DP} = (\text{high side} - \text{low side})$$

$$\text{DP} = (\text{pressure of liquid head} + \text{vapor pressure}) - \text{vapor pressure} \tag{7.2}$$

Note: 1 ft. $H_2O = 0.433$ psig at 60 °F.

The DP should be read in hydrostatic pressure units such as inches of water column. Multiply the DP by the specific gravity of the fluid to get the fluid level above the tap.

7.3.2 Capacitance

A capacitor consists of two plates electrically isolated from each other by a dielectric (nonconductor). The plates have an area (A) and are separated by a distance (D). Different mediums have a specific dielectric constant (K). This is illustrated in (Equation 7.3).

$$C = K\frac{A}{D} \tag{7.3}$$

For capacitance level measurement, a capacitor is formed when a level-sensing electrode is installed in a vessel. The metal rod of the electrode acts as one plate of the capacitor, and the metal tank wall acts as the other plate (Figure 7.5). If the tank wall is not metallic, a reference electrode is used. Because the plate area and gap distance remain constant, capacitance varies as a function of the dielectric constant of the substance filling the gap.

As the level rises, the air or gas around the electrode is displaced by the process material, which has a different dielectric constant. Therefore, there is a corresponding change in capacitance between the electrode and the vessel wall. The measured capacitance is proportional to the level. Capacitance probes can be used for liquids or solids in granular or powder form.

7.3.3 Ultrasonic

Level is measured using the ultrasonic method by transmitting a sound pulse from the instrument to the surface of a material (Figure 7.6). The sound pulse, traveling at the speed of sound, is reflected off the surface, and the echo returns to the instrument where it is detected. A shorter time for the

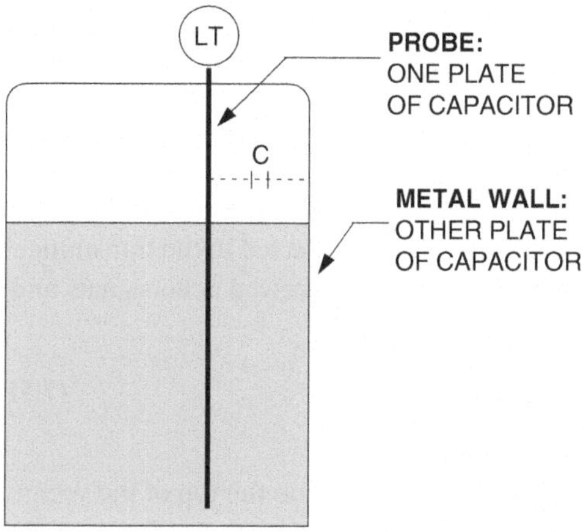

Figure 7.5 Capacitance level probe. C, capacitance; LT, level transmitter.

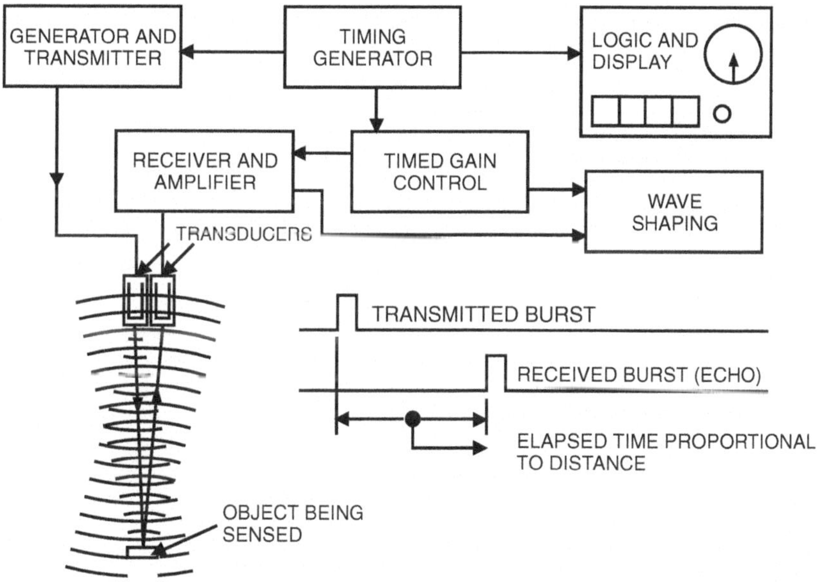

Figure 7.6 Ultrasonic measurement.

transmitted signal to return means the level is closer to the sensor and, therefore, higher in the tank. The distance to the surface is proportional to the travel time (Equation 7.4).

$$\text{distance} = \text{speed of sound} \times \frac{\text{time}}{2} \tag{7.4}$$

Ultrasonic transmitters can be installed on the top of the vessel, measuring down to the surface through the vapor space, or at the bottom of the vessel, measuring up to the surface through the liquid.

Ultrasonic transmitters in a top-down arrangement can measure liquid or solid levels.

Ultrasonic transmitters depend on the speed of sound being constant in the space between the transducer and the surface to be measured. In a pressurized vessel or tank where the vapor space pressure is not constant, the speed of sound will not be constant, which can cause more errors in the level measurement. Dusty space, such as the tops of grain silos, can also cause errors for ultrasonic transmitters.

7.3.4 Radar

A low-energy pulse of microwaves, traveling at the speed of light, is transmitted. When the pulse reaches the product level, a significant part of the microwave energy is reflected to the transmitter. The transmitter measures the time delay between the transmitted and received echo signals and calculates the distance to the liquid surface (Equation 7.5).

$$\text{distance} = \text{speed of light} \times \frac{\text{time}}{2} \tag{7.5}$$

Radar is used with liquids, slurries, sludges, and some solids.

In a guided wave radar (GWR) installation, the transmitter is mounted on the top of the vessel, and the antenna probe extends into the vessel beyond the normal product level. Antennas might be rigid or flexible.

Noncontacting radar level is like GWR, but there is no antenna probe; the antenna is mounted on the top of the vessel and does not contact the product. See Figure 7.7.

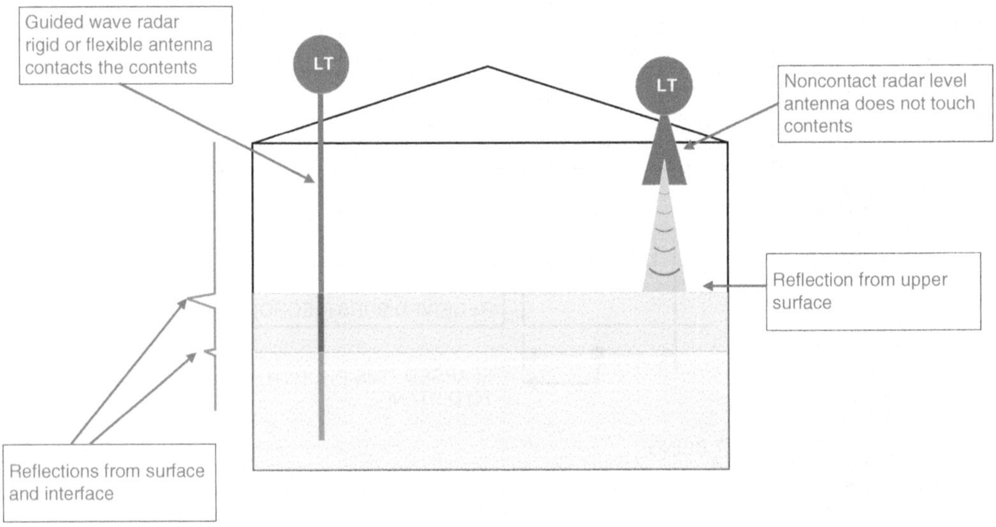

Figure 7.7 Radar level measurement.

With GWR, because some of the microwave energy will continue down the probe beyond the first fluid surface, a second echo can be detected from an interface between two liquids at a point below the initial liquid level. Therefore, GWR is often used for measuring liquid and liquid interfaces.

7.4 Level Measurement Uncertainty

CCST Level III

Tolerance is determined by application. For many applications, such as level control, repeatability and reliability of measurement might be more important than high accuracy. The highest accuracy would be required in systems such as tank gauging systems used for custody measurement or inventory.

It is not always as practical to calculate a total probable error (TPE) for level instruments as it is for temperature or pressure. The manufacturer's reference accuracy indicates whether a device can meet the required tolerance. Measuring accuracy can be influenced by the installation and conditions within the tank or vessel more than the instrument. Ambient and process temperatures affect the analog output.

7.4.1 Top-Down Level Measurement Errors

Top-down measurements can be affected by product turbulence from filling, agitation, or mixing (ripples). Agitation and turbulence cause the surface level to change. Uneven surfaces in a solids storage silo will affect top-down measurements.

Ultrasonic signals are affected by changes in the speed of sound in the vapor space. The speed of sound can change with the temperature and pressure of the vapor.

Radar signals can be scattered by agitation on the surface and by foam and emulsion. Changes in pressure, temperature, and most vapor space conditions have no impact on the accuracy of radar measurements.

7.4.2 Capacitance Level Measurement Errors

Capacitance depends on a stable dielectric of the medium throughout the height of the level column, so the overall accuracy of capacitance level measurement will be impacted if the dielectric constant changes. Buildup on capacitance probes will also introduce errors.

7.4.3 Hydrostatic Level Measurement Errors

Calculating the TPE of a DP for level is like calculating TPE for a pressure application with a few added considerations. The base accuracy of DP instruments is provided by the manufacturer.

Density (thus, specific gravity) will change with the temperature. Some systems correct for this by measuring temperature or by directly measuring density. The temperature might not be constant throughout the whole depth of the fluid, so the density might not be constant. The accuracy of the temperature or density instrument should be included in the TPE of the corrected level calculation.

When using DP with capillaries, consider the temperature effect on the density of the fill fluid. The manufacturer's literature will have information on this. Some manufacturer literature will have charts of TPE versus temperature for the assembled system.

When a remote sensor (dual sensors) is used with DP, the TPE calculation should include the accuracy of both sensors.

7.5 Level Calibration

7.5.1 Introduction

Every type of level device requires different calibration procedures. Many devices have no practical means of field calibration other than checking them against the actual level, which must be manually measured.

Indirect level measurements can sometimes be calibrated in the field by simulating the measured variable (such as DP). Many direct level measurements generally require changing the level in the container.

As with other types of devices, just causing the transmitter to simulate an output will only check the calibration of the instrument's analog output electronics and the programmable logic controller (PLC) analog input. It does not check the sensor. Some smart instruments have built-in diagnostics that can detect anomalies or issues with the electronics and maybe the sensors.

Sometimes, a functional check is all that is required when repeatability and reliability are more important than accuracy. For example, to confirm that an ultrasound gauge is working, point the ultrasound at an object about 10 feet away, and see if it reads 10 feet.

7.5.2 Level Calibration Equipment: Gauge Tapes, Bobs, and Bars

Calibrating level equipment sometimes requires measuring the tank level manually and changing the level to different points. Graduated tapes or bars (Figure 7.8) are tools used to measure the level in a tank manually.

There are two methods of obtaining a gauge reading: innage and ullage (outage). The reference gauge height from the reference point to the tank bottom or datum plate must be known. See Figure 7.9.

Innage gauging is a direct measurement of liquid depth. The tape, with the weighted bob on the end, is lowered until it just touches the bottom of the tank or the datum plate. (A *datum plate* is a plate installed near the bottom of the tank at a known distance from the gauge hatch.) Cut paste is applied to the tape as it is lowered. The cut paste changes colors when it is in the liquid, indicating the liquid level on the graduated tape. Innage gauging can only be used when the tank bottom or datum plate is clean of sediment and debris.

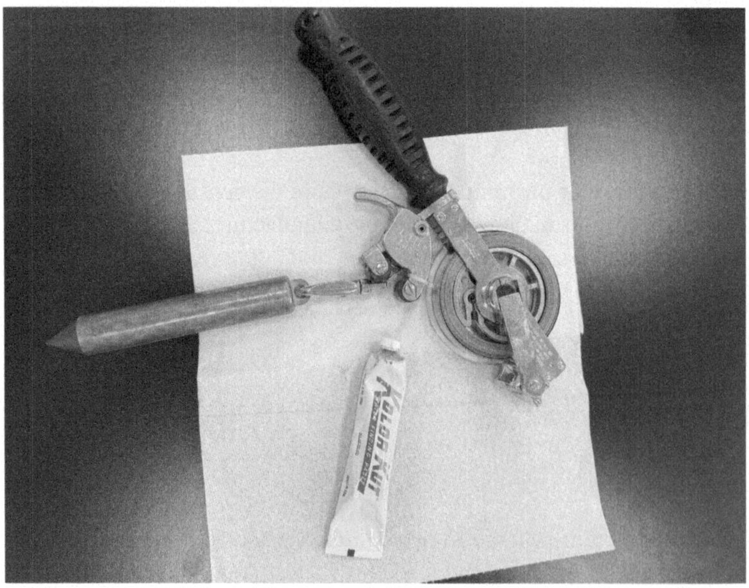

Figure 7.8 Typical gauge tape with innage bob and cut paste.

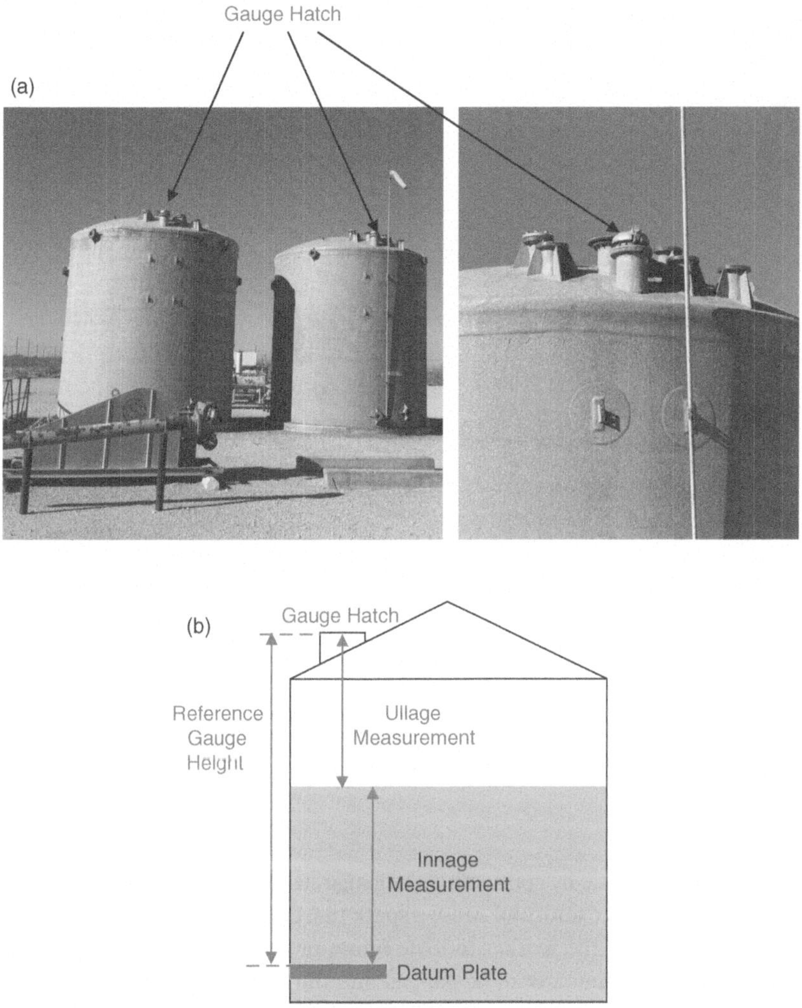

Figure 7.9 Gauge hatches and gauging diagram.

Ullage gauging is an indirect method. Cut paste is applied to the bob. The bob is lowered until a small portion is in the liquid. The graduated tape is read at the reference point. Innage can be calculated by subtracting the ullage gauge reading from the reference gauge height.

A good reference for manual tank gauging is the API *Manual of Petroleum Measurement Standards*, Chapter 3.1A – "Standard Practice for the Manual Gauging of Petroleum and Petroleum Products" [1].

There are manual, non-electronic (Figure 7.8), and electronic types of gauges (Figure 7.10).

As with other types of calibration standards, a working tape used for measurements must be verified periodically against a reference device (master tape) following a standardized procedure. The reference master tape must be certified as traceable to the national measurement standards authority.

Low-pressure and atmospheric tanks usually have a hatch that can be opened for the gauging procedure. Pressurized tanks or tanks with toxic fumes might have isolation valves for gauging (Figure 7.10).

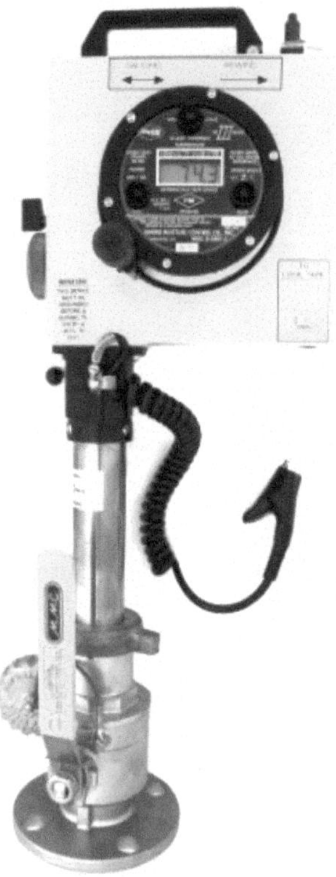

Figure 7.10 Portable electronic gauging tape with isolation valve [2]. Reproduced with permission of MMC International Corporation.

7.5.3 Safety

Gauging tapes must be grounded for static by connecting the gauge strap to the tank. Static electricity can build up as the gauge is run in the tank and cause a spark that can ignite flammable vapors.

Opening the gauge hatch can expose the worker to toxic fumes or release flammable vapor. It is important to understand what is in the tank to be gauged and the associated hazards. Review the Safety Data Sheet (SDS) for the materials in the tank. Personal protective equipment might be required.

Tank gauging often involves working at heights, so precautions should be considered if falling is a risk.

7.5.4 Calibrating a DP Level Transmitter

7.5.4.1 Procedure

If a DP transmitter used for level does not have remote seals, capillaries, or remote sensors, a DP level calibration procedure could resemble a pressure transmitter procedure, but determining the range depends on the application.

Where diaphragm seals, remote seals, capillaries, or remote sensors are used, it might be challenging to apply pressure for verification unless block valves are installed on the taps for that purpose. In the case where the sensors can be isolated from the process, follow these steps:

1) Isolate the top and bottom.
2) Vent the top to the atmosphere (low side).

3) Connect the pressure source to the bottom (high side).
4) Carry out a typical pressure calibration over the span of the device using the same method as that used for a gauge pressure transmitter.

7.5.4.2 Suppressed/Elevated Zero

If the DP cell is not located at an elevation that corresponds to 0% level in the tank, it must be calibrated to account for the difference in elevation because of head pressure in the sensing leg. This calibration adjustment is called *zero elevation* when the cell is located above the lower tap; it is called *zero suppression* when the cell is located below the lower tap (Figure 7.11).

The pressure on the low side is the same as the pressure in the vapor space, which compensates for the pressure that the vapor space adds to the liquid hydrostatic pressure on the high side.

To calculate the amount of elevation or suppression, the distance (D) between the zero reference level and the high-pressure port of the DP cell must be accurately measured.

In this example, the low-side connection is called a *dry leg* because the sensing line is not filled with liquid. It only senses the pressure in the vapor space. The disadvantage of a dry leg is that condensation can accumulate in the tubing and the liquid level in the tubing will exert pressure on the low side, which introduces an offset in the pressure measurement.

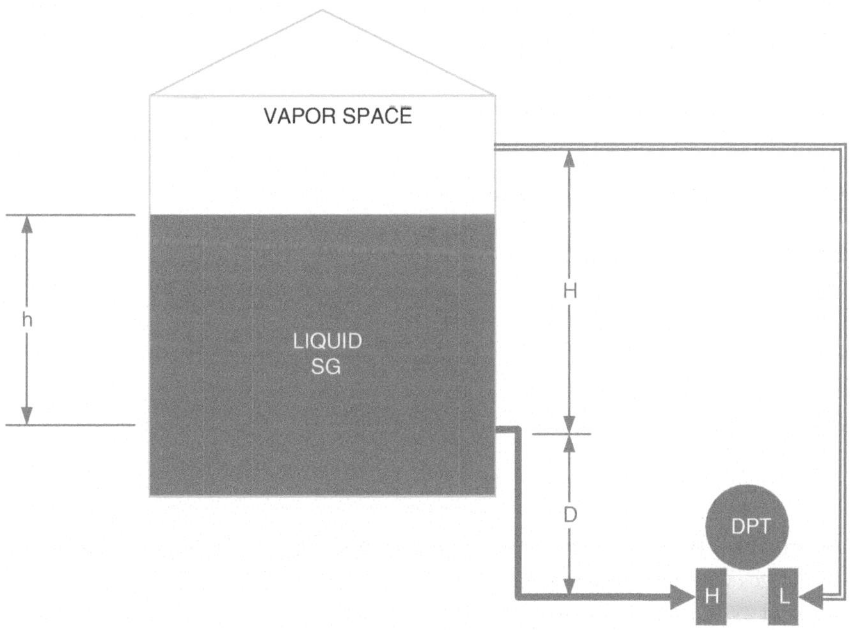

$$DP = P_{high\ side} - P_{low\ side}$$
$$LRV\ (Zero) = D \times SG$$
$$URV = (D + H) \times SG$$
$$Span = URV - LRV = H \times SG$$

Figure 7.11 Suppressed zero example. D, distance from the level of the lower tap to the level of the DP cell; DPT, differential pressure transmitter; h, height of the liquid level above the lower tap; H, height of the upper tap above the lower tap; H, high side of the DPT sensor; L, low side of the DPT sensor; SG, specific gravity.

7.5.4.3 Specific Gravity

Specific gravity is the ratio of the density of a process material to the density of water or air. It is particularly important to consider the specific gravity of the process material when calibrating DP level devices. If the specific gravity is assumed to be 1.0, or not considered at all, a significant error will be introduced if the process material is not water or has a specific gravity of 1.0.

The unit of measure, inches of water, is a pressure measurement, just like pounds per square inch. It does not represent the actual inches of level. The density of the fluid will affect the pressure exerted by a column of fluid.

For example, a 100 in. column of oil with a specific gravity of 0.8 will result in a pressure of 80 in. H_2O.

You will usually see inches of water abbreviated in. WC (inches water column) or in. H_2O (inches water).

Water has a density of 62.4 lbm/ft.^{3} at 4 °C. To determine the specific gravity of a process material, divide the material density by the standard density of water. For example, an oil with a density of 54.8 lbm/ft.^{3} has a specific gravity of 0.88 (Equation 7.6).

$$\frac{54.8\ \text{lbm/ft.}^3}{62.4\ \text{lbm/ft.}^3} = 0.88 \tag{7.6}$$

To properly calibrate a DP transmitter for a sensor monitoring the level of this oil, the nominal test points for input pressure must be multiplied by 0.88. Specifically, if the transmitter is calibrated for level 0 in. to 100 in. = 4–20 mA output, use the information in Table 7.4 to properly adjust the transmitter for the process fluid specific gravity of 0.88.

Table 7.4 Differential pressure level calibration example test points.

Nominal (in.)	Actual Test Pressure (in. H_2O) NOM. × 0.88	Expected Output (mA)
10	8.8	5.60
25	22.0	8.00
50	44.0	12.00
75	66.0	16.00
100	88.0	20.00

NOM, nominal.

It is important to remember that fluid densities change with the temperature. Figure 7.12 shows how water density changes with the temperature.

The density of water at 39 °F is typically the reference used for calculating the specific gravity of another liquid. For level measurement, we usually use the density of the process fluid at the normal operating temperature and divide it by the density of water at 39 °F.

To convert pressure units of psi to units of in. H_2O, multiply psi by 27.72 in. H_2O/psi.

7.5.4.4 Dry and Wet Legs

For DP measurements, the pressure of the vapor space in pressurized vessels is connected to the low-pressure side of the DP cell. When process vapors are noncondensing, noncorrosive, and nonplugging, the leg external to the vessel that is connected to the DP cell is a *dry leg*. If any

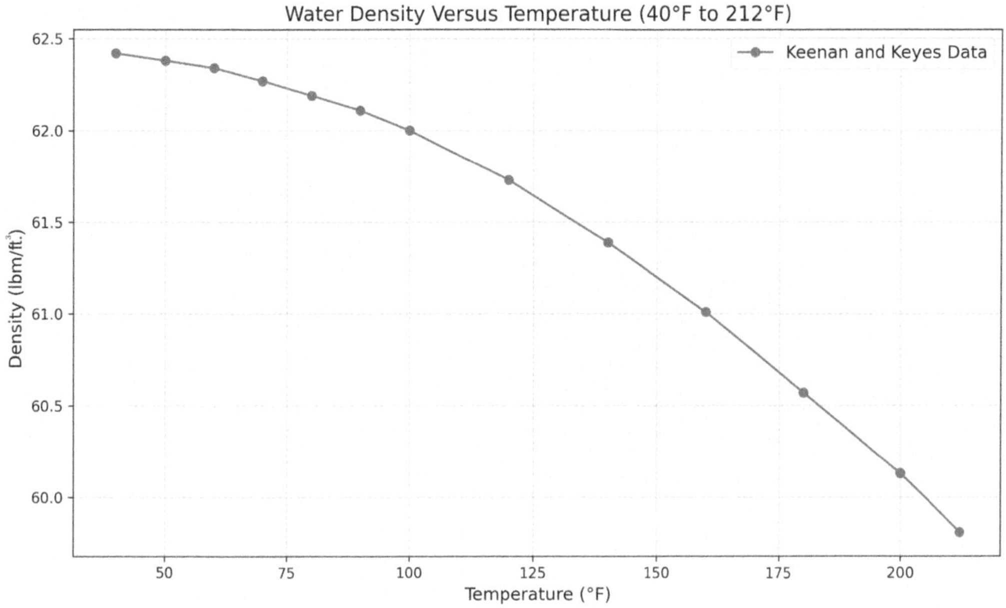

Figure 7.12 Water density versus temperature.

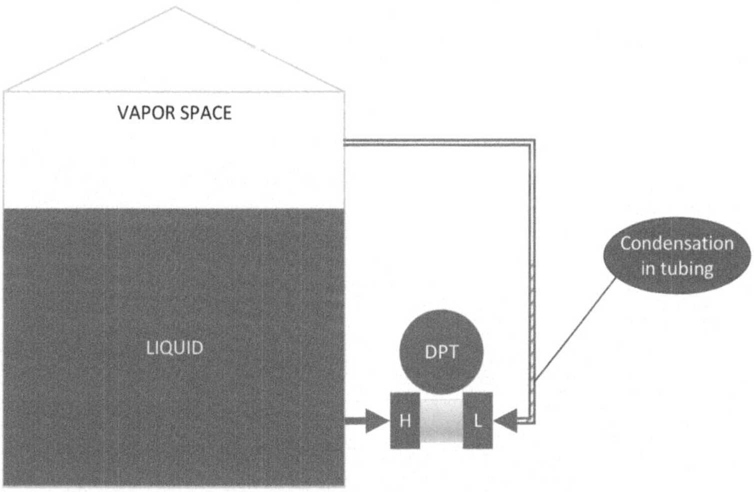

Figure 7.13 Condensation in a dry leg. DPT, differential pressure transmitter; H, high side of the DPT sensor; L, low side of the DPT sensor.

condensate accumulates in this leg, a significant error will result in the level indication. If there is condensate in the dry leg, it adds pressure to the low-pressure side of the DP cell (Figure 7.13). This would reduce the DP sensed and result in a level indication or output that is lower than the actual level.

To prevent condensation from affecting the measurement, the reference leg can be filled to form a wet leg (Figure 7.14). The specific gravity and height of the reference column must be accurately determined and accounted for in any cell calibration.

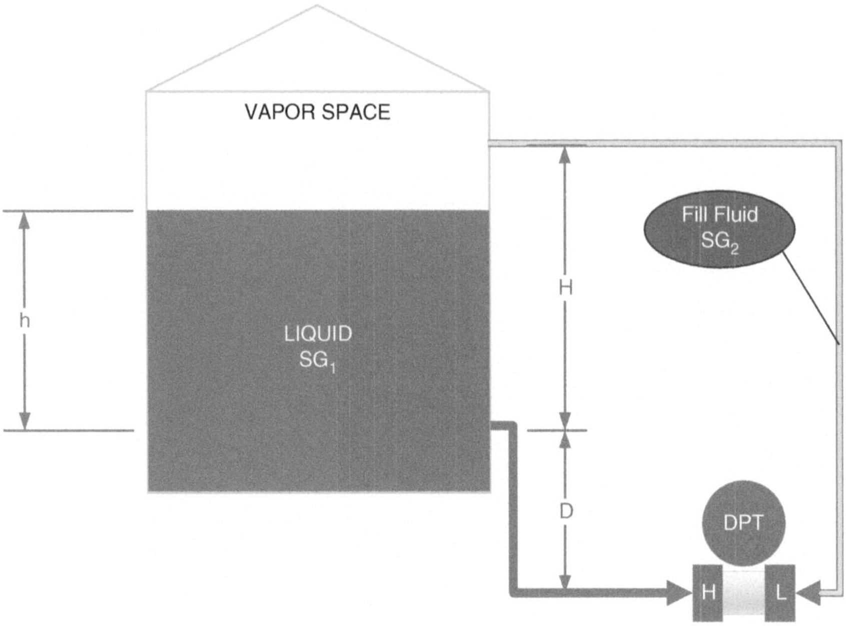

Figure 7.14 Wet leg. DPT, differential pressure transmitter; H, high side of the DPT sensor; L, low side of the DPT sensor; SG, specific gravity.

7.5.4.5 Diaphragm Seals and Filled Capillary Tubes

Diaphragm seal systems are used to separate the process from the device (Figure 7.15). These are often used in level applications.

These prevent sediment and solids from plugging the ports and eliminate the issues associated with dry legs and wet legs. These systems are filled with an incompressible fluid and sealed at the factory, so there is no chance of compressible gas forming.

Figure 7.15 Sealed membranes/diaphragms [3]. Rosemount DP Level Transmitters and 1199 Diaphragm Seal Systems, product data sheet 00813-0100-4016, Rev UC. Reproduced courtesy of Emerson, www.emerson.com.

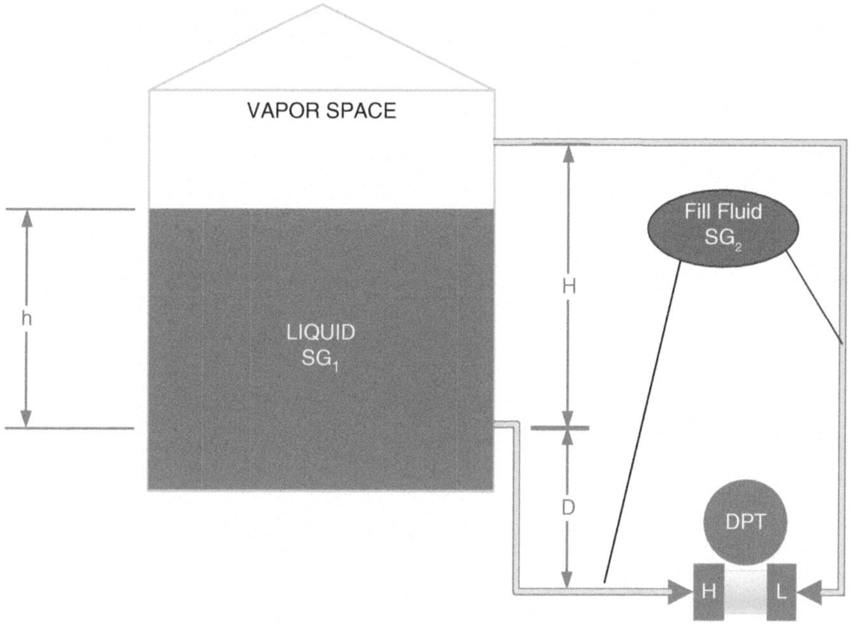

Figure 7.16 Installation with filled capillaries. D, distance from the level of the lower tap to the level of the DP cell; DPT, differential pressure transmitter; h, height of the liquid level above the lower tap; H, height of the upper tap above the lower tap; H, high side of the DPT sensor; L, low side of the DPT sensor; SG, specific gravity.

A small amount of extra uncertainty can be introduced by temperature effects on the specific gravity of the fill fluid and fill fluid thermal expansion. The manufacturer's literature will provide details. If the capillary lengths are the same on the low and high sides, this can minimize the temperature effects because the effect is the same on both sides.

Calculating the upper range value (URV) and lower range value (LRV) for these would work the same as for the wet leg example (Figure 7.14), except the low side and high side are both filled with the same fluid of the same specific gravity. See Figure 7.16. Some manufacturers make this easy with smart transmitters assembled for level applications; given the installation measurements, the configuration software will make the calculations.

7.5.4.6 Dual Sensors, Remote Sensors, and Multiple Transmitters

DP level systems with two sensors connected to a single transmitter are available (Figure 7.17). One sensor goes on the lower connection of the tank or vessel, and the other goes to the upper connection. The transmitter will provide an output proportional to the level that is calculated in the transmitter by subtracting the upper tap pressure from the lower tap pressure.

7.5.4.7 Interface Measurement with DP

For the purposes of level instrumentation, *interface* typically refers to the point at which two non-mixable fluids meet, such as oil and water. Special considerations must be made to accurately measure the level of each fluid and/or the level of the interface.

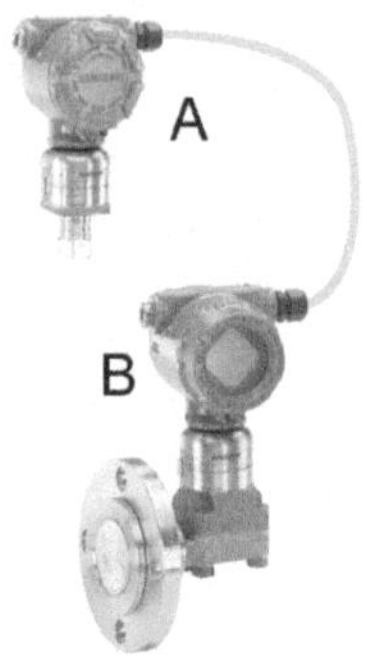

Figure 7.17 Dual-sensor, hydrostatic level transmitter [3]. Rosemount DP Level Transmitters and 1199 Diaphragm Seal Systems, product data sheet 00813-0100-4016, Rev UC. Reproduced courtesy of Emerson, www.emerson.com.

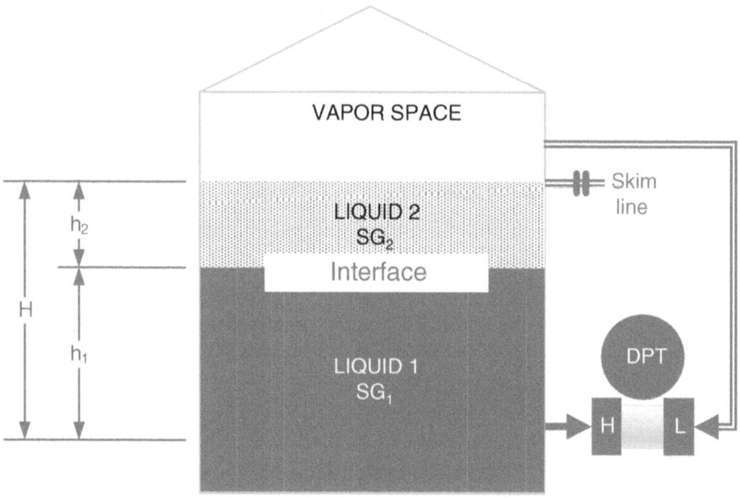

Figure 7.18 Differential pressure measurement of the interface level. D, distance from the level of the lower tap to the level of the DP cell; DPT, differential pressure transmitter; h, height of the liquid level above the lower tap; H, height of the upper tap above the lower tap; H, high side of the DPT sensor; L, low side of the DPT sensor; SG, specific gravity.

An example of a skim tank is shown in Figure 7.18. The total liquid height is fixed by the skim line where fluid flows out of the tank. The lighter oil (Liquid 2) floats on top of the water (Liquid 1) because it has a lower specific gravity. The DP transmitter can be used to measure the level of the interface between the oil and water (h_1).

It would be configured this way:

- Configure the LRV so 4 mA represents the tank full of oil ($h_1 = 0$)
- Configure the URV so 20 mA represents the tank full of water ($h_1 = H$)

$$\text{LRV (Zero)} = (H \times SG_2)$$

$$\text{URV} = (H \times SG_1)$$

$$\text{Span} = \text{URV} - \text{LRV} = H \times (SG_1 - SG_2)$$

7.5.5 Calibrating a Capacitance Level Instrument

The initial calibration of a capacitance level instrument following installation is vendor-specific. It usually involves adjusting the actual tank level to one point, accurately measuring that level, and entering the actual level in the calibration setup (or pushing a button at the zero level). The tank level is then adjusted to a second point, accurately measuring that level and entering the actual level in the calibration setup (or pushing a button at the 100% level). The calibration points must be accurately determined and should be as far apart as possible.

Periodic calibrations can follow a similar procedure. If readings are incorrect, inspect the capacitance probe. Contamination buildup or soiling can form on the probe. A high level of material buildup can affect the measurement result. Check it again after cleaning. If the level instrument requires adjustment, follow the instructions in the manufacturer's instrument technical manual.

7.5.6 GWR

Radar level usually requires the manufacturer's software to look at the reflected waveforms to verify and troubleshoot.

One method is to have a verification reflector permanently attached to the antenna:

> The reflector is used to test and continuously verify that the transmitter functions properly in tank and chamber/pipe installations. Compared to traditional diagnostics that only monitor the transmitter electronics, the reflector can also be used to diagnose the upper parts of the probe inside the tank for buildup, corrosion monitoring, and other process-related conditions. The primary use cases for the reflector are:
>
> - Verification of transmitter and probe (proof-testing)
> - High-level supervision (continuous monitoring of high-level conditions) ([4], Section 5.3)

If the radar is removed from the vessel, a clamp can be placed at the measured distance along the antenna to verify that a reflection is received by the transmitter and reads the correct distance.

7.5.7 Calibrating an Ultrasonic Level Instrument

There are two methods for calibrating an ultrasonic level transmitter. Ideally, the instrument can be removed, and the cabling has enough slack to move the instrument to an open area with an unobstructed path to a smooth surface (such as a wall or floor). If this is the case, follow Method 1. If this is not the case, follow Method 2.

7.5.7.1 Method 1

1) Determine the distance from the face of the ultrasonic transducer to the 100% level. This should already be documented as part of the instrument setup. For this example, we are going to assume the distance is 12 in.
2) Determine the test point distances. Realize that the distance from the test surface simulates the distance from the top of the liquid level and takes into account the distance determined in Step 1. For example, if the distance from the transducer face to the 100% level is 12 in. and the tank level indication is 0 in. to 100 in., use the test points in Table 7.5.
3) Turn off the transducer.
4) Use appropriate safety precautions and remove the instrument from the top of the tank.
5) Use a tape measure to mark off the distances from the test surface at each desired test point determined in Step 2.
6) Place the instrument at a distance equal to the first test point from the wall or floor that will be used to bounce the signal back (the test surface simulates the top of the liquid level).
7) Turn the transducer on and record the level reading.
8) Move to the next distance and record the level reading. Repeat until all as-found readings have been obtained.

Table 7.5 Ultrasonic level test points.

Span (%)	Desired Level Indication (in.)	Distance from Test Surface (in.)
0	0	112
25	25	87
50	50	62
75	75	37
100	100	12

9) If necessary, adjust the instrument according to the instructions in the manufacturer's instrument technical manual.
10) If adjustments were made, repeat Steps 6 through 8 to obtain as-left readings.
11) Return the instrument to operational condition.

7.5.7.2 Method 2

1) Use manual tank gauging to determine the actual liquid level in the tank.
2) Adjust the level in the tank to each of the desired set points throughout the range of the instrument and record the reading from the ultrasonic instrument at each test point.
3) If necessary, adjust the instrument according to the instructions in the manufacturer's instrument technical manual.
4) If adjustments were made, repeat Step 2 of this method to obtain as-left readings.
5) Return the system to operational condition.

7.5.8 Calibrating a Level Loop Using Manual Gauging

In this example, a gauge tape is used to manually find the actual level of the fluid in the tank and compare it with the reading in the automation system for calibration. If a gauge tape can be used on the tank or vessel, this method will work for calibrating any level measurement technology (e.g., servo, radar, hydrostatic, capacitance, and ultrasonic).

If this manual measurement method is intended to be accurate enough for custody transfer measurement, the gauging tape used for the procedure must be calibrated against a higher reference standard.

See the example level loop calibration procedure in Appendix D. The procedure in this example is for innage gauging (measuring the depth of the liquid), where the tape is lowered all the way to the tank bottom or datum plate and the actual level is indicated on the tape.

Review Questions

7.1 Hydrostatic pressure level measurement is an example of:
a) Indirect measurement
b) Direct measurement

7.2 If hydrostatic pressure is used for level measurement in a pressurized vessel, then why must the vessel pressure be measured in addition to the hydrostatic pressure at the bottom?

a) To calculate the volume of fluid in the tank from the level
b) To compensate for the pressure exerted by the gas vapor space above the liquid
c) To correct for sensor drift over time
d) To determine the density of the process fluid

7.3 Why is it important to consider the effect of the temperature on the capillary fill fluid in a remotely sealed hydrostatic level transmitter?
a) High temperatures cause the capillary fluid to dissolve into the process fluid.
b) Fill fluid evaporates at high temperatures.
c) It is not important.
d) Temperature changes can expand or contract the fill fluid, affecting the pressure reading on the DP transmitter and introducing error.

7.4 How can temperature affect the level of water in a vessel?
a) Temperature has no effect on the level of water in a vessel.
b) As the temperature increases, water expands, raising the actual level in the vessel.
c) As the temperature increases, water density decreases and causes the level to drop.

7.5 Why is a manual tape gauge often used to calibrate a level loop?
a) It directly measures the pressure in the vessel.
b) It is the only method approved by international standards.
c) It directly measures the actual level in the vessel for comparison to the loop measurement.
d) It automatically logs data for historical records.

CCST Level III

7.6 A radar level transmitter has these specifications:
- Reference accuracy: ±0.12 in. (at 77 °F)
- Repeatability: ±0.04 in.
- Ambient temperature effect: ±0.0044 in. per degree Fahrenheit from reference temperature

a) What is the TPE of the radar device at an ambient temperature of 50 °F?
b) The transmitter is set to measure a range of 0 in. to 120 in. The transmitter 4–20 mA output is connected to a PLC input that has an uncertainty of ±0.1% of the range. What is the TPE of the temperature measurement loop?

References

Note: The dates and editions for standards and other references in this chapter are current as of publication. The reader is encouraged to check for later revisions or editions.

1 American Petroleum Institute [API], 2013, *Standard Practice for the Manual Gauging of Petroleum and Petroleum Products*, Chapter 3.1A – "Standard Practice for the Manual Gauging of Petroleum and Petroleum Products." Washington, DC: API.

2 Marine Moisture Control (MMC) International, 2025, *Portable Electronic Gauging Tapes.* MMC International. https://mmcintl.com/project/portable-electronic-gauging-tapes/. Accessed February 1, 2025.

3 Emerson, 2023, March, Rosemount DP Level Transmitters and 1199 Diaphragm Seal Systems Product Data Sheet 00813–0100-4016, Rev UC. Clayton, MS: Emerson.

4 Emerson, 2015, March, *Rosemount 5300 Series Reference Manual 00809–1600-4530, Rev AA, Commissioning and Validation Procedures for Rosemount Radar*. Shakopee, MN: Rosemount/Emerson.

8

Flow Instrument Calibration

8.1 What Is Flow?

Flow refers to the movement of a substance (liquid, gas, or granular material) from one point to another. Flowmeters are used to measure low-viscosity gases like hydrogen, high-viscosity fluids like peanut butter, and everything in between. They can also be used to measure the flow of slurries, powders, grains, and other materials.

According to ANSI/ISA 51-1-1979 (R1993), *Process Instrumentation Terminology*, a flowmeter measures the flow rate or quantity of a fluid moving in a pipe, duct, channel, or other conduit [1]. Flowmeters usually have two or three parts:

- The primary device (sensor) produces a signal with a defined relationship to the fluid flow. The primary device might be in contact with the fluid or external.
- The secondary device converts the signal from the primary device to a display or output signal that represents the flow or quantity and can be used for indication or control.
- A tertiary device performs more flow computations, corrections, and totalizations and perhaps stores a history of flow quantity and meter configuration.

8.2 Flow Terms and Units of Measure

Flow rate is usually given in terms of quantity per unit of time. Frequently used units of measure for flow are listed in Table 8.1. Flow rates can be provided in terms of volume or mass per unit of time.

The volumetric rate can be given at actual flowing conditions of temperature and pressure, or it might be converted to represent the flow rate at other specified conditions. In Table 8.1, *cfd* refers to the flow rate of a gas at flowing conditions; *scfd* refers to the flow rate with the volume modified to a standardized temperature and pressure.

Why is volumetric flow sometimes corrected to a standardized set of base conditions? Consider 1 ft.^{3} of methane in the pipeline at 90 °F and 120 psia. This same chunk of methane at base conditions of 60 °F and 14.696 psia would occupy 7.8 ft.^{3}; a gas company customer will be billed based on the standard cubic feet, not the volume at flowing conditions. Conditions in the pipeline change over time. The pipe will be warmer on a summer day than on a cold winter night. Correcting to standardized conditions allows the company to always bill on the same basis. Base conditions of 60 °F and 14.696 psia are frequently used in the natural gas industry in the United States.

Table 8.1 Units of measure for flow.

SI Standard Unit of Measure	
Cubic meter per second (volume rate)	m^3/s
Kilogram per second (mass rate)	kg/s
Other Common Units of Measure	
Cubic centimeters per minute (typical for samples to analyzers)	cc/min.
Liters per minute	L/min.
Pounds (mass) per hour	lbm/hr.
Gallons per hour	gph or gal./hr.
Barrels per day (common for liquid)	bpd or bbl/d
Cubic feet per day (common for gas)	cfd or $ft.^3/d$
Standard cubic feet per day (gas)	scfd

Gas volumes vary greatly with changes in temperature and pressure. Liquid volumes also change with temperature, but they change very little with pressure. Corrections to a standardized set of conditions can be made in the flowmeter secondary and tertiary devices or in higher-level computing systems.

8.3 Types of Flow Instruments

There are many sensor technologies for flow measurement. The sensor type used in a particular application depends on many factors, including the process fluid measured, pressure, temperature, allowable pressure drop, density, conductivity, viscosity, pipe size and orientation, flow rate and/or flow total required, accuracy tolerance requirements, control system interface, accessibility, and maintenance requirements.

Flowmeters measure flow in one of three main ways:

1) Measure velocity
2) Measure volume flow
3) Measure mass flow

A control system technician (CST) might encounter many flowmeter technologies (Table 8.2). This chapter briefly describes a few sensor types:

- Differential pressure (restriction)
- Magnetic
- Vortex
- Turbine
- Ultrasonic
- Coriolis
- Thermal mass

Flow measurement instruments have a primary flow element (also called a *flow sensor* or *primary flow device*) and a secondary device (flow transmitter and/or flow computer). The flow element is in

Table 8.2 Flow measurement technologies.

Differential Pressure	Velocity	The flow rate is proportional to the square root of the pressure drop across a restriction.
Positive Displacement	Volume	The flowmeter repeatedly captures fixed quantities of the fluid, counts the number of times these quantities are displaced, and divides the flow into discrete measurable volumes.
Turbine	Velocity	Flow across turbine blades (like fan blades) causes the turbine to spin. The rate of rotation is proportional to the velocity of the flow.
Magnetic	Velocity	When a liquid conductor moves in a pipe, it will induce a voltage proportional to the flow velocity (based on Faraday's principle of induction).
Doppler Ultrasonic	Velocity	The frequency of an ultrasonic signal changes when it reflects off particles moving with a flowing fluid. The frequency shift is proportional to the velocity of the fluid.
Transit Ultrasonic	Velocity	The flowmeter measures the difference in transit time between ultrasonic signals sent upstream and downstream through a flowing fluid. The difference in travel times is proportional to the fluid velocity.
Vortex	Velocity	A bluff body causes alternating vortices to form on either side. The frequency of vortices is proportional to the fluid velocity.
Thermal Mass	Mass	Heat is drawn away from a heated sensor at a rate proportional to the mass flow rate of the fluid.
Coriolis	Mass	Fluid moving through a vibrating tube causes twisting forces on the flow tube. The twist is related to the mass flow rate (Coriolis principle).

the process and produces a signal. The transmitter converts the signal to a standard signal that can be read by a control system. The transmitters are unique to the type of flow element. The transmitter and element can be calibrated together or separately.

A flow computer might also be involved. A flow computer is usually used for custody transfer and other fiduciary meter applications; it keeps an auditable history of raw measurements and configuration changes.

8.3.1 Differential Pressure

Restriction-type flow instruments are based on the principle that the flow rate is proportional to the square root of the differential pressure across the restriction divided by the density at flowing conditions (Equation 8.1).

$$Q \sim \sqrt{\frac{\Delta P}{\rho}} \tag{8.1}$$

The primary flow element in Figure 8.1 is an orifice plate. This restriction creates a pressure drop that can be converted to the flow rate. An alternate method would be to install another type of differential pressure (DP) primary element—such as a wedge meter, Venturi flowmeter, flow nozzle, cone meter, or pitot tube—in line with the process flow to restrict the liquid flow. Examples are shown in Figure 8.2.

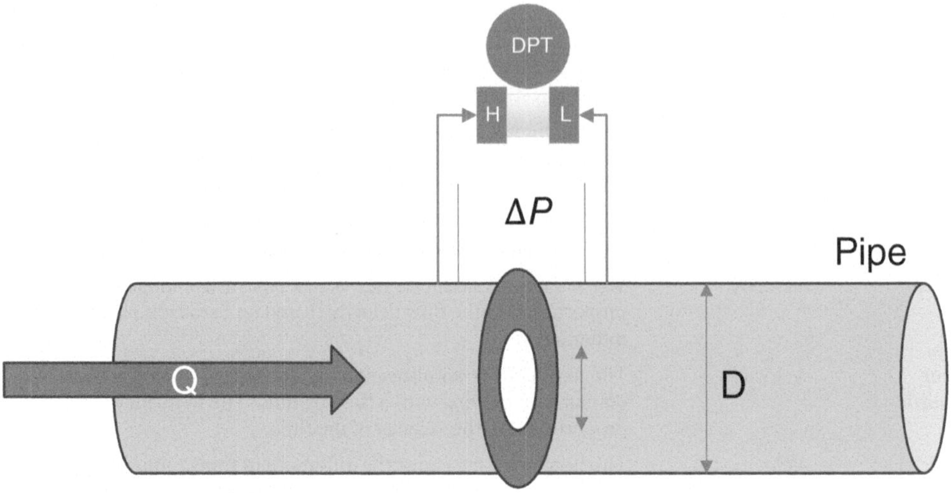

Q = volumetric flow rate
P = flowing/static pressure in the line
ΔP = DP across the restriction
ρ = density at flowing conditions
D = pipe ID
d = restriction ID

Figure 8.1 DP flow measurement – orifice plate.

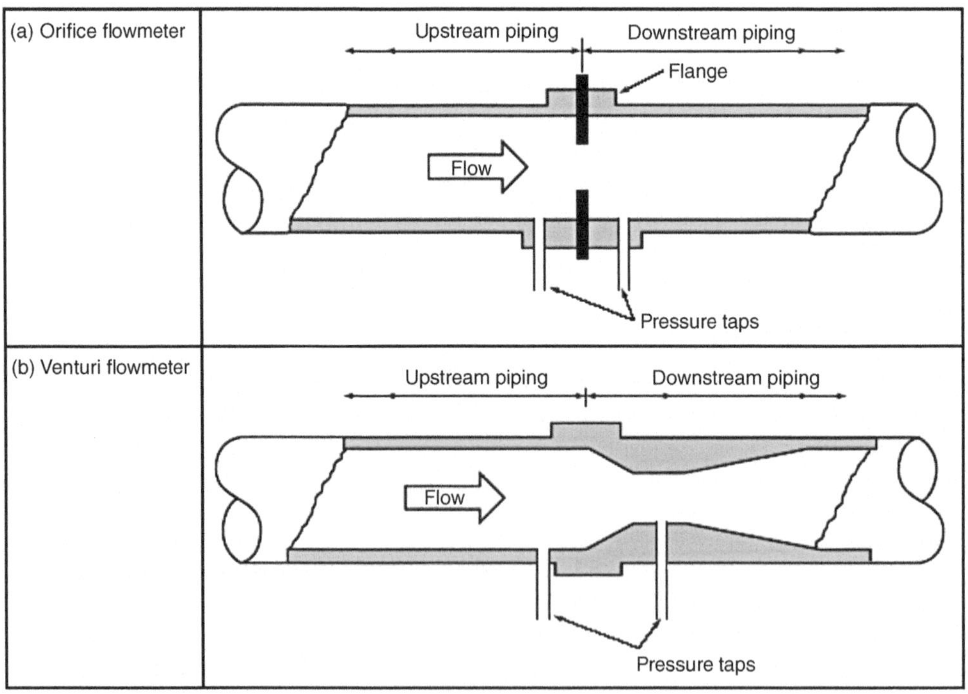

Figure 8.2 DP flow elements. H, high; L, low. API, *Manual of Petroleum Measurement Standards*, Chapter 22 – "Testing Protocol Section 2 Differential Pressure Flow Measurement Devices." [2]. Reproduced courtesy of the American Petroleum Institute (API).

(c) Flow nozzle

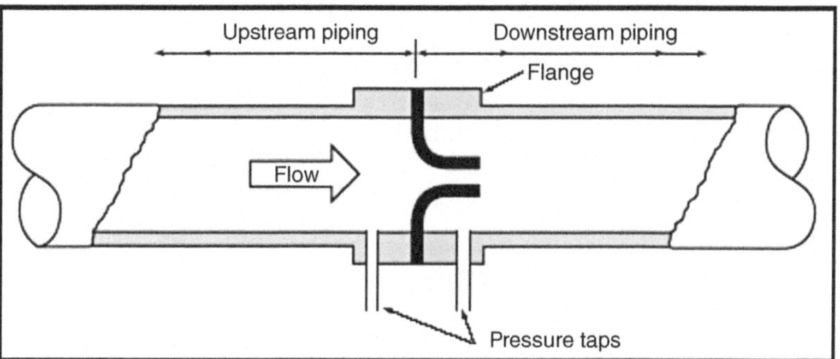

(d) Cone meter

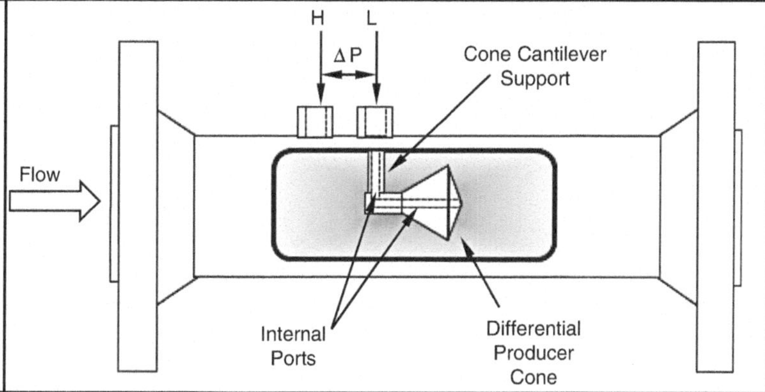

(e) Wedge meter

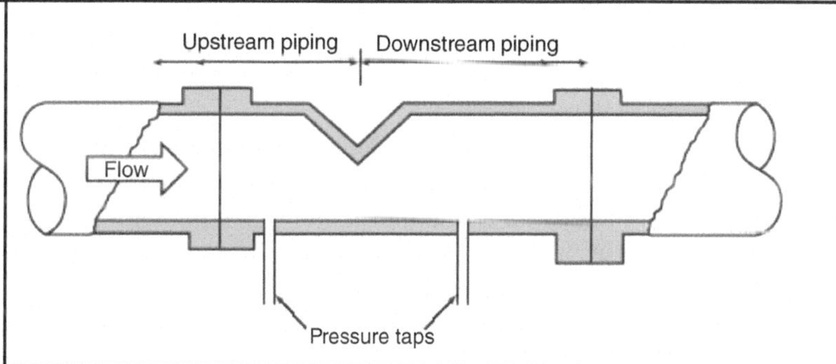

(f) Pitot tube

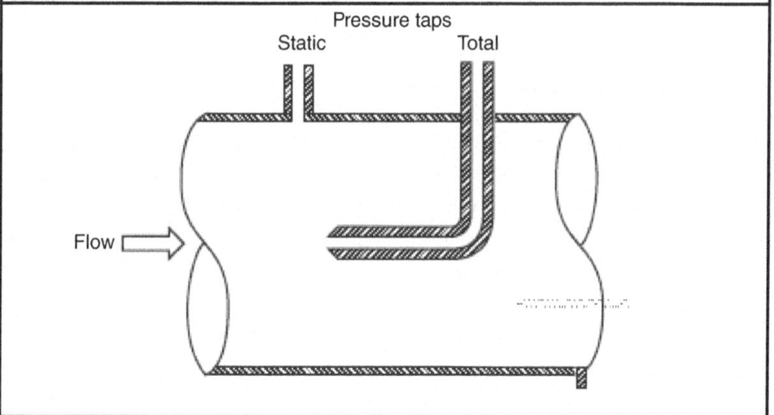

Figure 8.2 (*Continued*)

(g) Multi-port averaging pitot tube

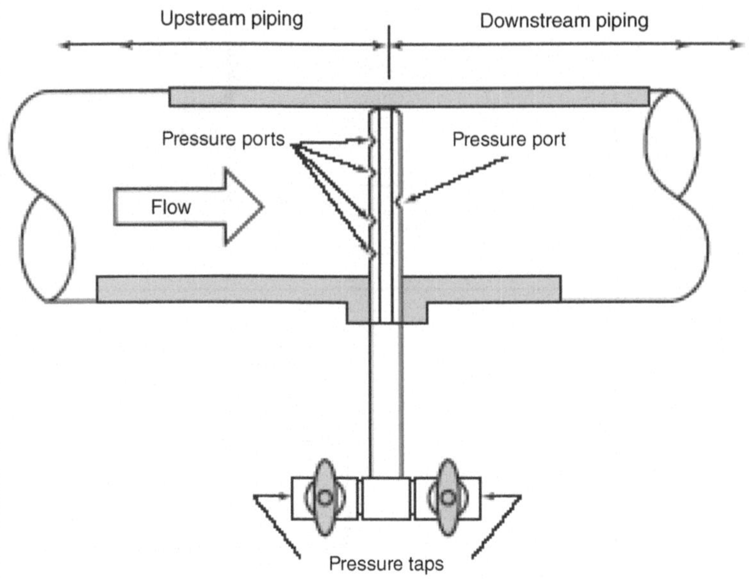

Figure 8.2 (*Continued*)

The secondary devices necessary for calculating the flow rate include DP, static pressure, fluid temperature, density, and a device to calculate the flow. Secondary devices can be electronic or mechanical.

The upstream pressure is connected to the high-pressure port of a DP transmitter. The downstream side is connected to the low-pressure port of a DP transmitter.

The square root is accounted for in the transmitter signal processing, the control system, or a flow computer. Mechanical chart recorders might use square root extractor springs in the DP cell. However, some older models use a separate square root extractor installed in the output signal loop. In this case, the installed square root extractor must also be calibrated for the loop to perform within specification.

8.3.2 Magnetic Flowmeter

The magnetic flowmeter (Figure 8.3) is a velocity-type meter. It is based on Faraday's law of electromagnetic induction. When a conductive liquid moves in a pipe having diameter D and travels with an average velocity (V) through a magnetic field (B), it will induce a voltage (E) that is proportional to the velocity of the fluid (Equation 8.2). The calibration factor (K) is given by the manufacturer for each device.

$$E = K \times V \tag{8.2}$$

Magnetic flowmeters can only be used to measure the flow of conductive liquids. As conductive liquid passes through the magnetic field, an electrical potential is induced. The change in potential varies directly with the liquid velocity. This voltage is sensed by electrodes in the walls of the flowmeter. The velocity (V) is calculated from the voltage, then the flow rate (Q) can be

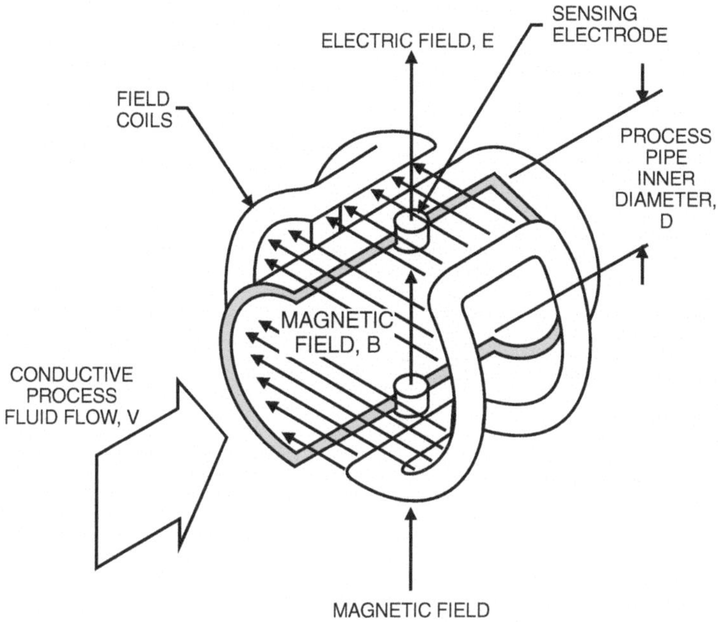

Figure 8.3 Magnetic flowmeter. B, magnetic field; D, diameter; E, voltage; V, velocity.

calculated based on the velocity and the cross-section area of the pipe, which is calculated from the diameter (D).

$$Q = V \times \pi \left(\frac{D}{2}\right)^2 \tag{8.3}$$

8.3.3 Ultrasonic Flowmeter

There are two types of ultrasonic flowmeters: Doppler (Figure 8.4) and transit time (Figure 8.5).

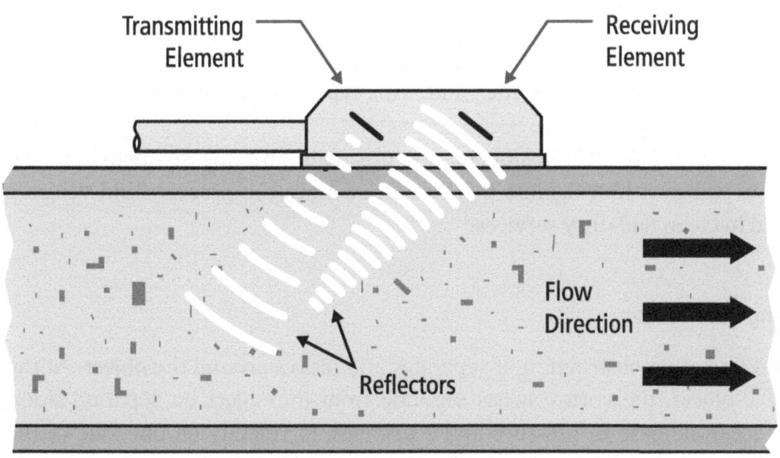

Figure 8.4 Doppler ultrasonic flowmeter.

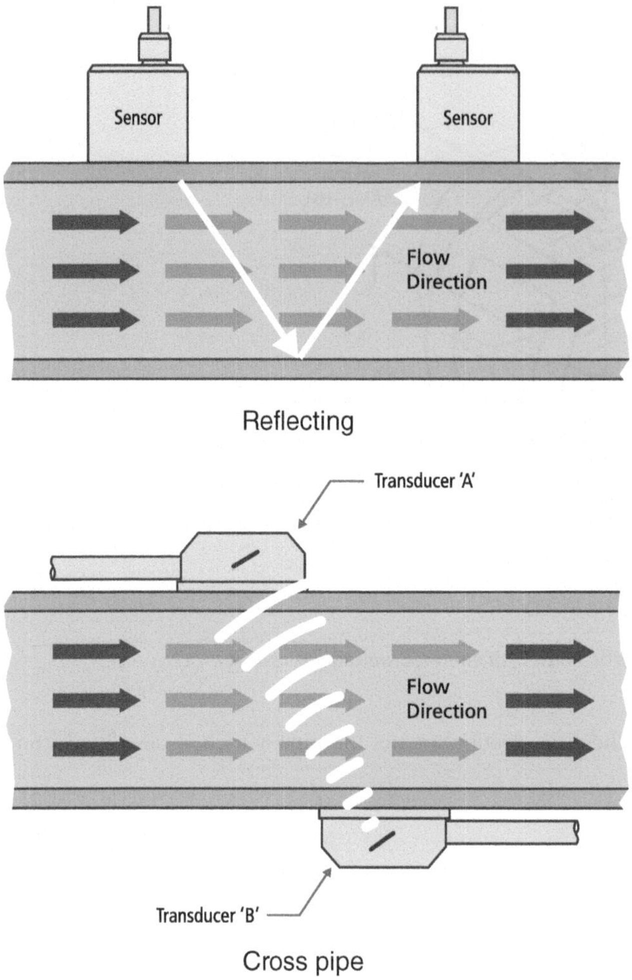

Figure 8.5 Transit time ultrasonic flowmeter.

The Doppler flow element works by measuring the shift in frequency of an ultrasonic wave sent by a transducer and reflected by impurities in the fluid. This type of meter only works in dirty service.

The time-of-flight meter has a pair of transducers, one upstream and one downstream. The difference in elapsed time going with or against the flow determines the process fluid velocity. This type of meter is used in clean and dirty services.

8.3.4 Vortex-Shedding Flowmeter

When a flowing medium strikes an obstruction, it separates and moves around the object. At the point of contact with the object, the vortex swirls separate from the object on separate sides. This shedding causes a local increase in pressure and a decrease in velocity on one side of the object. A local decrease in pressure with a corresponding increase in velocity occurs on the other side of the object. After shedding from one side, the process is reversed. The frequency of shedding reversal is proportional to the velocity of the fluid passing the obstruction.

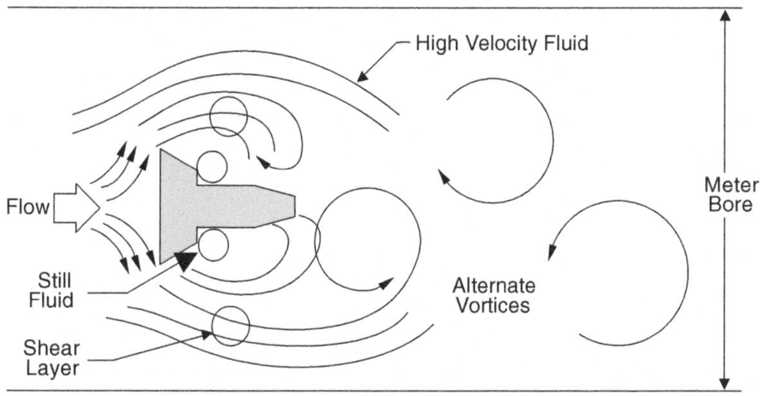

Figure 8.6 Vortex-shedding flowmeter.

Picture a flag waving in the wind. The flagpole (bluff body) creates vortex swirls as the wind blows past. The alternating vortices can be seen as waves in the flag.

A vortex-shedding flowmeter (Figure 8.6) has three basic parts: a bluff body (or shedder bar), a sensor, and a transmitter. In the process line, flowing liquid strikes the bluff body and vortices are shed alternately on each side of the bluff body. The flow velocity determines the frequency at which the vortices are shed. Sensors built into the bluff body detect the shedding of the vortices on alternate sides of the bluff body.

These sensors can be force sensors or temperature sensors (thermistors). The sensor then converts the energy created by the vortices into electrical pulses. These pulses are conditioned and sent to the transmitter, which generates the output signal proportional to the flow rate. The volumetric flow rate (Q) is the vortex-shedding frequency (f) multiplied by the meter coefficient (K).

8.3.5 Turbine Flowmeter

The turbine flowmeter (Figure 8.7) is a velocity-type meter. It is a mechanical flowmeter that measures flow by means of a spinning turbine or rotor using an arrangement of moving parts. The multibladed rotor is mounted at right angles to the flow, suspended in the process fluid stream on a free-running bearing. The rotation speed is proportional to the flow velocity and, thus,

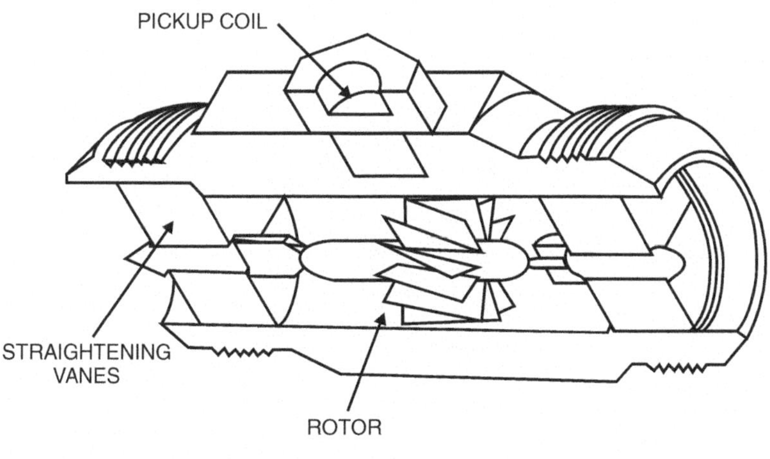

Figure 8.7 Turbine flowmeter.

the volumetric flow rate. Solid-state devices or mechanical sensors detect the turbine rotation. As shown in Figure 8.7, a pickup coil is used to detect the rotary speed for signal processing. As each blade passes the coil, a voltage pulse is generated.

Each pulse represents a discrete volume. The number of pulses per unit of volume is the meter's K-factor. The output is a continuous sine wave with a frequency proportional to the flow rate.

8.3.6 Coriolis Mass Flowmeter

A mass flowmeter measures the flow rate in mass per unit time rather than volume. A Coriolis meter (Figure 8.8) will directly measure both mass flow and density.

Fluid moving through a vibrating tube causes twisting forces on the flow tube (Figure 8.9). A Coriolis flowmeter has two of these tubes in parallel. Electrical pulses are applied to an exciter coil in the middle of the flow tubes to cause the tubes to move back and forth in opposite directions. Magnetic sensors placed on each corner of the tubes sense the movement.

The meter or transmitter electronics will calculate the mass flow rate and density from the phase shift and frequency (Figure 8.10). Outputs for mass rate and density are usually available from a Coriolis meter.

- Each sensor outputs a sine wave representing the movement of that sensor.
- The frequency of the sine wave is proportional to the density of the fluid in the tube.
- If the flow rate is zero, then there is no twist in the coil, so the two waves are in phase. If there is flow, then the tube twists, so the sine waves are out of phase.
- The time difference between the two waves is proportional to the mass flow rate.

The Coriolis primary device is the meter tubes with the associated electronics. A secondary device will be a transmitter that converts the signal from the primary meter to an analog or digital output to the control system.

Figure 8.8 Coriolis flowmeters. Sergey Ryzhov. Adobe Stock Photos.

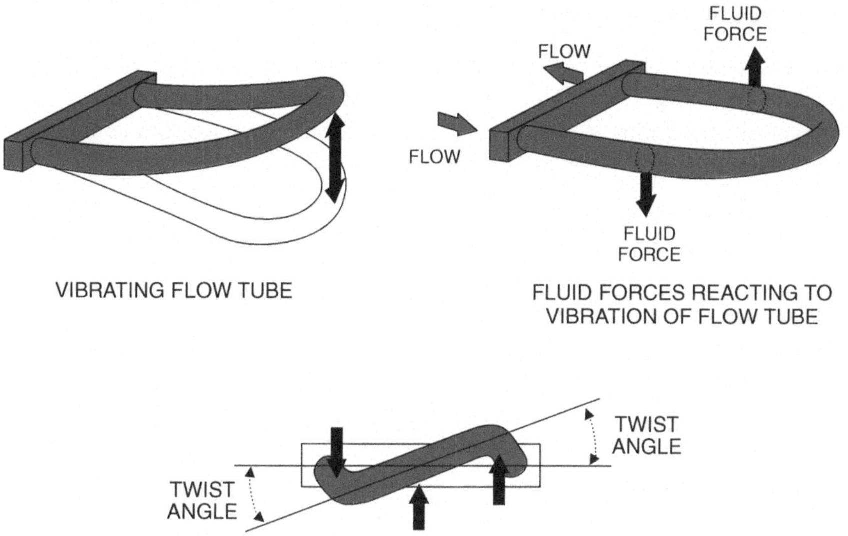

Figure 8.9 Coriolis effect twisting the meter tubes.

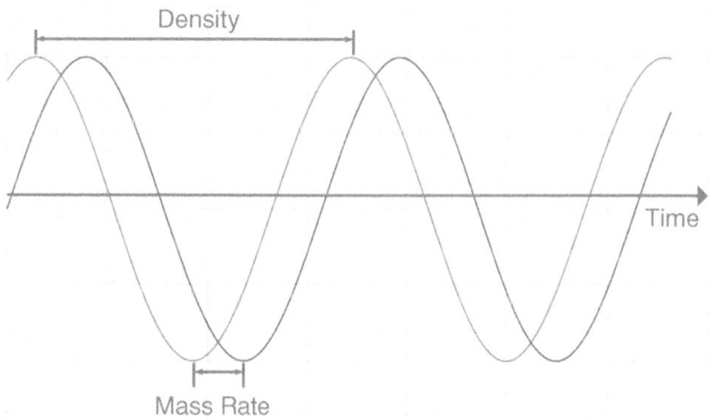

Figure 8.10 Coriolis meter tube pickup signals.

8.3.7 Thermal Mass Flowmeter

This measuring principle is that heat is drawn from a heated body when a fluid flows past. A thermal flowmeter contains two temperature sensors for this purpose. One sensor measures the current fluid temperature as a reference. The second sensor is heated so that it has a constant temperature differential relative to the reference sensor.

As soon as the fluid begins to flow in the measuring tube, the temperature sensors cool off due to the fluid flowing past; the higher the flow, the greater the cooling effect. The electric current required to maintain the temperature differential is thus a direct measure of mass flow. The mass flow rate is proportional to the electric current going to the heating element.

Thermal mass flowmeters are often used in ductwork for heating, ventilation, and air-conditioning systems.

8.3.8 Electronic Gas Measurement System

An electronic gas measurement (EGM) system includes the secondary and tertiary devices necessary to measure and record gas flow. An EGM can be used for any type of primary device. Primary devices are typically transmitters, and the tertiary device is usually a gas flow computer.

For example, an EGM for an orifice meter would include:

- Primary device: Orifice plate flow element
- Secondary devices:
  - o DP transmitter
  - o Static pressure transmitter
  - o Temperature transmitter
- Tertiary device: Flow computer

8.4 Flow Measurement Uncertainty

CCST Level III

8.4.1 Flow Measurement Criticality, Tolerance, and Calibration Frequency

Tolerance is established based on the application. For example, custody measurement tolerance will be established in the custody contract. Regulatory requirements might specify tolerance and frequency for some applications, such as flare meters for environmental monitoring or measurements of hydrocarbons on federal lands. These categories should be included in the calibration program's criticality classification method.

Meter criticality might be determined by the volume of product exchanged (Table 8.3) and the associated financial or product quality risk.

Criticality might be lower with less stringent frequency and tolerance for surveillance and flow control applications. Operational meters might only be inspected when requested rather than on a schedule.

Frequency of proving can be specified on a time basis or on the amount of product that flowed through the meter. The frequency required for proving varies from several times a day to twice a year, or even longer, depending on the

- value of the liquid,
- cost-to-benefit to prove,

Table 8.3 Example calibration frequency versus volume.

Volume, Mscf/d	Frequency
Very-Low Volume, 0–35	N/A
Low Volume, 35–200	5 years
High Volume, 200–1000	2 years
Very-High Volume, 1000 and higher	1 year

mscf/d, thousand standard cubic feet per day; N/A, not applicable.

- meter proving history,
- meter system stability,
- variations of operating systems, and
- contractual, regulatory, or organizational requirements.

8.4.2 Uncertainty

Uncertainties depend on the technology. There will be errors associated with the primary element, secondary element, installation, process conditions, and environmental conditions.

Calculated uncertainty will be valid over a specified range of flow rates and other process conditions. In addition to the operating pressure and temperature, operating conditions can include the fluid state, Reynolds number, and speed of sound in the fluid.

If a meter measures multiple different fluids, such as various grades of crude oil coming into a refinery, uncertainty must be calculated for each fluid.

Some DP primary devices, such as wedges and cones, will have meter factors that are unique to the specific assembly. The meter factor is determined by individual calibration at the factory.

Orifice meter tubes and orifice plates are manufactured to tolerances specified in industry standards. They are machined, not assembled, unlike other DP devices. Orifice meter runs manufactured to meet the standard tolerance do not require factory calibration; therefore, orifice uncertainty can be calculated per industry standard.

8.4.3 Example: Uncertainty of an Orifice Meter

Many factors influence the overall measurement uncertainty associated with an orifice meter. Major contributors include construction tolerances in the primary meter components, tolerances of empirical coefficient of discharge databases or in situ flow calibrations, predictability of and variations in the fluid's physical properties, and uncertainties associated with the secondary devices, such as pressure and temperature sensors [3].

Estimating the uncertainty of an orifice meter installation can be quite involved, so this text will not go into the details of determining the uncertainty of each factor. Consult the standards for detailed information. This example demonstrates that it is not sufficient to consider only the accuracy of the DP, pressure (P), and temperature (T) instruments in the overall meter uncertainty.

The example shown in Equation 8.3 is a summary of an example calculation from API MPMS 14.3.1 ([3], AGA 3 Part 1). The table refers to figures and other references within the standard.

This uncertainty estimate is for this gas mass flow using Equation 8.4 [3]:

$$q_m = C_d E_v Y \left(\pi/4\right) d^2 \sqrt{2g_c \frac{G_i Mr_{air} P_f}{Z_f R T_f} \Delta P} \tag{8.4}$$

The DP, pressure, and time uncertainties are the accuracies of those transmitters given by the manufacturer. In this example, the base accuracies from the literature are used, but the influences of the environment and process can be included, as shown in previous chapters.

The orifice diameter and pipe diameter uncertainties are determined as specified in the American Gas Association standard AGA 3 Part 2 [3]. If an online density instrument is used, use the base accuracy of the density device. The uncertainty of the other calculated factors is given in the standard (Figure 8.11).

		Uncertainty U_{95} (%)	Sensitivity Coefficient, *S*	$(U_{95}S)^2$
C_d	Discharge (Figure 4)	0.44	1	0.1936
Y	Expansion factor (Table 5)	0.02	1	0.0004
d	Orifice diameter [a]	0.05	$2/(1-\beta^4)$	0.0114
D	Pipe diameter [b]	0.25	$-2\beta^4/(1-\beta^4)$	0.0011
ΔP	Differential pressure	0.50	0.5	0.0625
P	Static pressure	0.50	0.5	0.0625
Z	Compressibility factor [c]	0.1	−0.5	0.0025
T	Temperature	0.25	−0.5	0.0156
G	Relative density	0.60	0.5	0.0900
Sum of squares				0.4396
Square root of sum of squares				0.6630

NOTE As the table shows, the overall gas flow measurement uncertainty at a 95 % confidence level is ±0.66 %.

a API *MPMS* Ch. 14.3, Part 2/AGA Report No. 3, Part 2/GPA 8185, Part 2–2000, Table 2-1.

b API *MPMS* Ch. 14.3, Part 2/AGA Report No. 3, Part 2/GPA 8185, Part 2–2000, Section 2.5.1.3.

c AGA Report No. 8.

Figure 8.11 Example uncertainty estimate for natural gas flow. API *Manual of Petroleum Measurement Standards*, Chapter 14.3.1 – "Orifice Metering of Natural Gas and Other Related Hydrocarbon Fluids – Concentric, Square-edged Orifice Meters – Part 1: General Equations and Uncertainty Guidelines," Table 7 [3]. Reproduced courtesy of the American Petroleum Institute (API).

8.4.4 Example: Uncertainty of a Coriolis Meter

Coriolis manufacturers provide flow accuracy and repeatability data. The accuracy includes the combined effects of repeatability, linearity, hysteresis, orientation, and other factors. The base performance specifications for Coriolis meters typically depend on the fluid. Coriolis meters are more accurate for liquids than for gases.

Meter accuracy is affected by pressure and temperature due to the effects on the stiffness of the tubes. The meter electronics or associated transmitter will adjust for these effects. Coriolis meters have temperature sensors on the tubes that can be used to adjust for the temperature effect. For pressure, a fixed value near the operating point can be used, or a pressure transmitter can be added.

Process temperature effect is defined as the change in sensor mass flow and density accuracy specification due to process temperature change away from the calibration temperature. The temperature effect on flow can be corrected by zeroing at the normal operating temperature.

The process pressure effect is defined as the change in sensor mass flow and density accuracy specification due to process pressure change away from the calibration pressure. This effect can be corrected by dynamic pressure input or a fixed meter factor.

Table 8.4 and Table 8.5 are specifications for a Micro Motion ELITE Coriolis flowmeter.

The process temperature effect is given as ±0.0005% of the maximum mass flow rate per degree Celsius. The process pressure effect is given as −0.0006% of the rate per psi.

When the flow rate approaches the low end of the flow range (less than 5% of the maximum rate), Coriolis meter accuracy begins to deviate from the stated accuracy rating (Figure 8.12).

Table 8.4 Coriolis accuracy and repeatability for liquids and slurries.

Performance Specification	Standard Option
Mass/volume flow accuracy	±0.10% of rate
Mass/volume flow repeatability	0.05% of rate
Density accuracy	±0.5 kg/m^3 (±0.0005 g/cm^3)
Density repeatability	0.2 kg/m^3 (0.0002 g/cm^3)

Table 8.5 Coriolis accuracy and repeatability for gases.

Performance Specification	Standard Models
Mass flow accuracy	±0.25% of rate
Mass flow repeatability	0.20% of rate
Mass flow linearity	±0.05% of rate up to a 0.2 Mach number

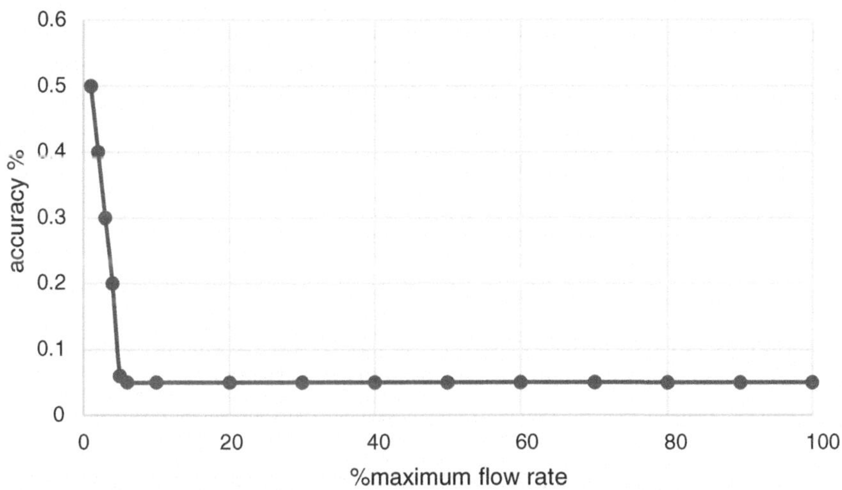

Figure 8.12 Coriolis accuracy versus flow rate.

8.5 Flowmeters Calibration

8.5.1 Introduction

Flowmeters are calibrated by comparison to a reference standard. The choice of reference standard and the method of calibrating a flowmeter depends on many things.

- **Fluid** – In most cases, reference standards and calibration methods for liquids differ from those for gases.
- **Technology** – Some flow technologies have specific calibration methods.
- **Criticality** – Higher criticality requires more stringent methods.
- **Application** – Calibration methods might be specified in regulations or custody contracts, or they might depend on product quality requirements.

- **Process** – It depends on whether the process flow can be interrupted for calibration.
- **Practicality** – Calibration can be performed in situ, or the meter can be removed from service and sent to a certified laboratory for calibration.

The reference standard is placed in series with the meter to calibrate during stable operating conditions. This in situ calibration is preferable over off-site calibration because it verifies the meter's performance using the exact product and process conditions the meter normally experiences during field operation. It also allows for a complete loop calibration rather than just calibrating the device. For in situ calibration with a reference standard, the calibration should be performed under conditions of pressure, temperature, and flow rate that are typical of the actual operation. There cannot be any leaks between the meters. The flow rate should be in a steady state.

However, in situ calibration is not always practical or possible. The meter can be removed from service and taken to a flow calibration facility for comparison with a reference standard. The meter that is in service is typically replaced by a calibrated meter to reduce the amount of time the process is interrupted. Once the meter is lab calibrated, it is put on the shelf, so it is available to be swapped again the next time a calibration is due.

8.5.2 Meter Proving

Two additional terms are associated with liquid flowmeters: *meter proving* and *meter factor*. Meter proving is the procedure used to determine a meter factor. The meter factor is the ratio of the prover volume measurement to the meter volume measurement (Equation 8.5). Both volumes must be corrected to standard conditions to calculate the meter factor.

$$\text{meter factor} = \frac{\text{volume measured in prover}}{\text{volume measured by meter}} \tag{8.5}$$

For mass flowmeters, the meter factor can be in terms of mass (Equation 8.6).

$$\text{meter factor} = \frac{\text{mass measured in prover}}{\text{mass measured by meter}} \tag{8.6}$$

This is especially important when the measurement is used for inventory or custody transactions of products. Once a meter factor is determined, it can be used in one of two ways:

1) Multiply the indicated volume from the meter by the meter factor to correct the indicated volume.
2) Adjust or replace the meter so it reads the same as the reference standard (a meter factor of 1). Organizations and contracts will usually specify a maximum acceptable meter factor. Also, if a meter factor changes significantly from one proving to the next, then the meter should be corrected or replaced.

Meter proving involves making multiple runs through the prover, which must agree within a specified tolerance. Meters should be proved at the typical operation conditions of pressure, temperature, and flow rate using the same liquid that usually flows in the line.

8.5.3 Flow Calibration Equipment

8.5.3.1 Gas Flow

For gas flow, there are three main methods for in situ calibration:

1) Compare the meter reading to a master meter. This method can be applied to any meter technology. Master meters can be sonic nozzles, orifice meters, or Coriolis meters.

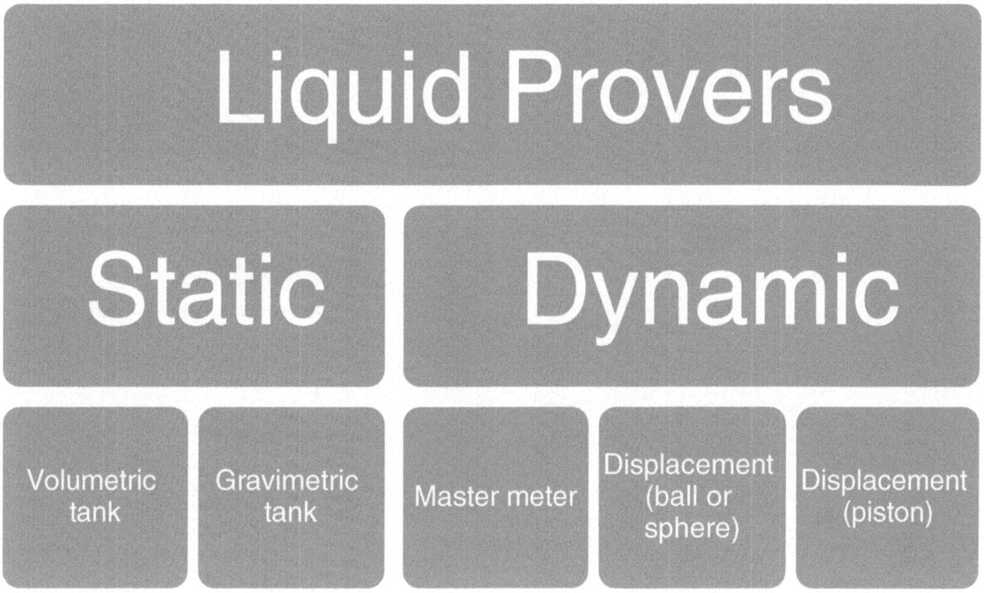

Figure 8.13 Liquid prover types.

2) Use the meter diagnostic available on some meters. Some Coriolis and ultrasonic models have diagnostics that can be used instead of comparing to a reference standard.
3) For orifice meters, calibrate the secondary devices and inspect the primary devices to confirm they are within tolerances.

8.5.3.2 Liquid Flow (Meter Provers)

Liquid meter reference standards for in situ calibration include master meter provers, volumetric provers, and gravimetric provers. The provers might be permanently installed for frequently calibrating multiple meters, or they might be portable, perhaps mounted on a trailer for transportability. See Figure 8.13.

8.5.3.3 Tank Provers

A tank prover is a calibrated vessel used to measure the volume of liquid passed through a meter. The known volume of the tank prover is compared with the meter-indicated volume to determine the meter factor or to calibrate the meter.

A volumetric tank prover can be an open or closed volumetric measure that generally has a graduated top neck and may have a graduated bottom neck. The volume is established between a shutoff valve or bottom-neck graduation and an upper-neck graduation. Tanks can be fixed (Figure 8.14) or portable (Figure 8.15) [4]. Components of a typical prover are shown in Figure 8.16. Some are very small (5 gal. or less), and others are large (more than 400 gal.). The required capacity of the prover tank will be determined based on the flow rate through the meter.

Tank provers incorporate a temperature sensor. If it is a closed tank, a pressure sensor is also required. The measured volumes from the meter and the prover should be corrected to standard pressure and temperature conditions for comparison.

Figure 8.14 Stationary volumetric meter prover [4]. Reproduced with permission of Meter Engineers.

Figure 8.15 Portable tank prover [4]. Reproduced with permission of Meter Engineers.

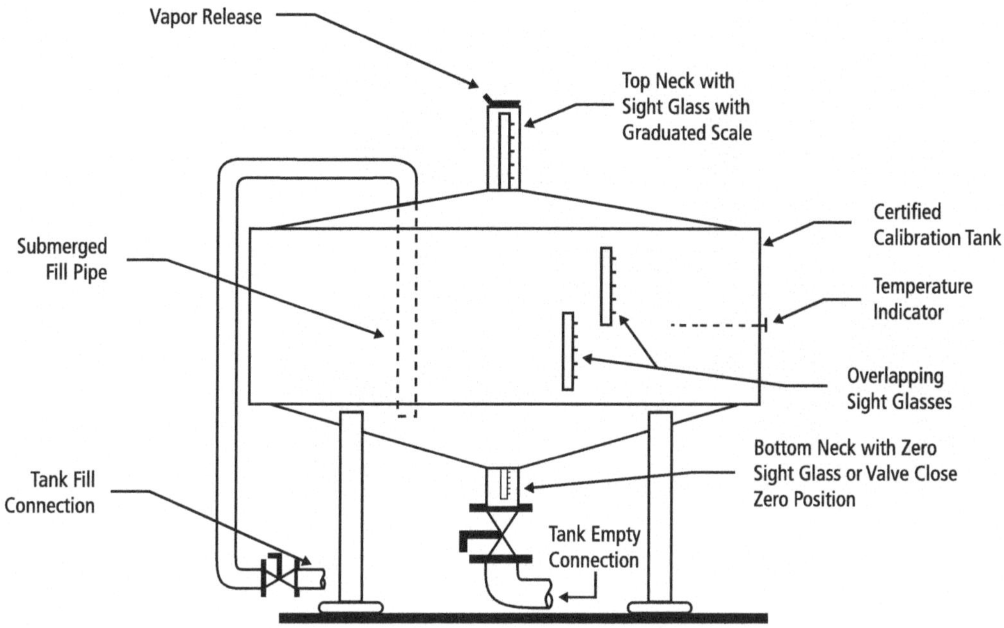

Figure 8.16 Components of a tank prover.

Proving tanks are calibrated against higher-level calibrated standards by one of two methods:

1) Fill the tank with water and then draw the water into a field standard tank that has been calibrated by a certified laboratory.
2) Calibrate using a master flowmeter.

After proving with a tank, the product can be pumped back into the system if facilities are available, or the tank is emptied into a disposal facility.

The tanks must be kept clean.

Gravimetric tank provers are similar to volumetric tank provers except that they work by weighing the fluid in the tank rather than measuring the volume. The mass and volume of the fluid can be determined from the weight and density of the fluid. Gravimetric provers are more suitable for a laboratory than a field environment, but portable systems exist, especially for small meters.

Master Meter Provers A master meter prover (Figure 8.17) uses a calibrated meter in series with the meter under test. The proving meter will be verified to a higher accuracy than the meter being calibrated (ideally, at least a 4:1 test uncertainty ratio [TUR]). Master meters are usually displacement, turbine, Coriolis, or ultrasonic meters.

The master meter volume registered is then compared to the test meter volume registered, and a new meter factor can be calculated from the totals.

Displacement Provers Displacement provers (Figure 8.18) have a calibrated section of pipe between two detectors. An oversized sphere or ball travels through the pipe as it is pushed by the process fluid. An accumulated volume reading from the meter under test is recorded when the displacer passes the first detector and again when it passes the second detector. The difference between

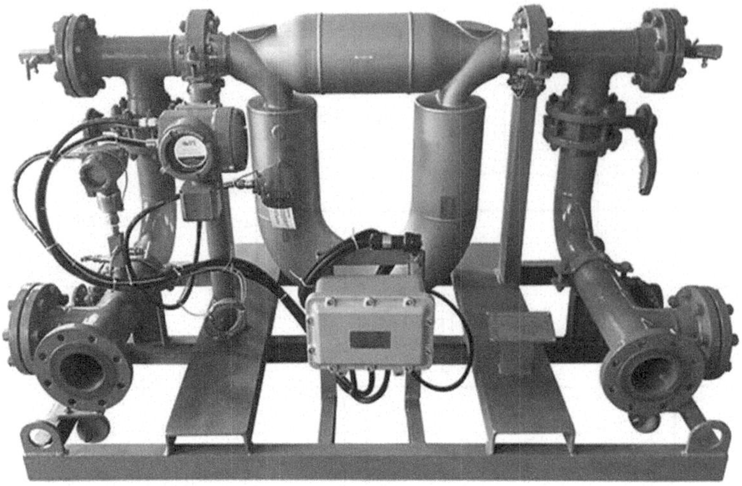

Figure 8.17 Master meter prover (Coriolis). Reproduced courtesy of Emerson, www.emerson.com.

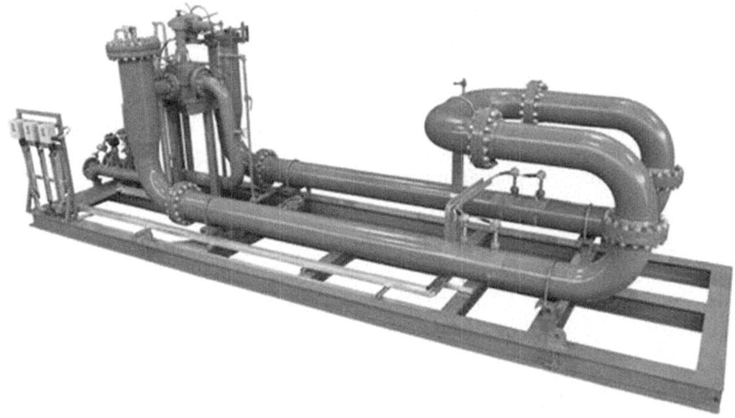

Figure 8.18 U-type meter prover. Reproduced courtesy of Emerson, www.emerson.com.

the opening and closing readings is compared to the known volume of the pipe between the two detectors.

The meter that is being proved is also connected to the proving counter. The counter is started and stopped when the displacing device actuates the two detectors at the ends of the calibrated section.

Displacement provers are usually automated to capture the starting and ending counts.

8.5.4 Safety

Many hazards must be considered when calibrating flowmeters. Operations must be informed and prepared. The CST must make preparations to avoid the hazards. A few ideas are listed in Table 8.6. Calibration procedures should list all potential hazards, and a job safety analysis should always be performed prior to doing a calibration in situ.

Always understand the hazards of potential exposure to the process fluids, such as pressure and temperature, corrosivity, and toxicity.

Table 8.6 Flow calibration hazards.

Blocking Flow to Check Zero Readings	This interrupts the process and causes back pressure. What happens when the flow is stopped? Identify where pressure relief points might expose the CST to process fluids.
Connecting Reference Standards in Series	Disconnecting piping can expose the CST to fluids. Ensure that isolation valves are working to avoid exposure to fluid under pressure.
Disposal of Fluids from Reference Standards	The CST should understand the hazards of exposure to the fluids and the safe handling procedures.
Portable Calibration Standards	Some reference standards are very heavy. Follow best practices to move and use them safely. Trailer-mounted equipment must be blocked to keep it from rolling.

CST, control system technician.

8.5.5 Examples

This chapter provides examples of custody transfer meters associated with crude oil or natural gas. These applications usually have calibration requirements specified in contracts, and the methods are based on industry standards. The examples are for field calibration, not laboratory calibration.

8.5.5.1 Example: Gas Orifice Meter and EGM

Consider an orifice meter with an EGM used to measure natural gas. Two standards apply:

- ANSI/API MPMS Ch. 14.3.1/AGA Report No. 3, API *Manual of Petroleum Measurement Standards*, Chapter 14.3 parts 1 through 4 – "Orifice Metering of Natural Gas and Other Related Hydrocarbon Fluids – Concentric, Square-edged Orifice Meters" [3]
- API *Manual of Petroleum Measurement Standards*, Chapter 21.1 – "Flow Measurement Using Electronic Metering Systems – Electronic Gas Measurement" [5]

Orifice meters that are manufactured and installed per the API 14.3 standard will all perform within an acceptable tolerance if the orifice plate and meter tube are clean, in good condition, and within the tolerances specified in the standard.

Likewise, if all the secondary devices (DP, pressure, and temperature transmitters) are performing within the tolerances specified for the installation, then the calculated flow rate will be within the tolerance. (See the example in Section 8.4.3.)

In lieu of calibrating or proving the meter against a reference flow standard, such as a master meter, the standards will accept verification that the primary and secondary devices are within tolerance. The procedure could be as follows:

1) Inspect the orifice plate for wear. The edge should be square and sharp. The orifice plate bore edge is considered too dull for accurate flow measurement if the upstream edge reflects a beam of light when viewed without magnification or if the upstream edge shows a beam of light when checked with an orifice edge gauge. The upstream and downstream edges of the orifice plate bore must be free from defects visible to the naked eye, such as flat spots, feathered texture, roughness, burrs, bumps, nicks, and notches.
2) If there is any doubt about whether the edge has sufficient quality for accurate metering, the orifice plate should be replaced.

3) Verify that the plate bore diameter and roundness are within specified tolerances.
4) Check the plate flatness.
5) Inspect the meter tube's inside surface for wear, corrosion, and roughness.
6) Check the inside diameter and roundness of the meter tube.
7) Calibrate the DP, pressure, and temperature transmitters as discussed in previous chapters. Ideally, perform a loop calibration, rather than a device calibration, on each to confirm the reading in the flow computer.

8.5.5.2 Example: Coriolis Meter for Natural Gas

This is an example of using meter diagnostics for verification instead of calibration against a reference standard. This example is based on the American Gas Association (AGA) Report 11 – "Measurement of Natural Gas by Coriolis Meter" [6].

Report 11 allows for performing a verification that the meter is operating properly and performing within tolerance [6]. Testing has shown that Coriolis meters free from flow tube corrosion, erosion, or mechanical damage perform as originally flow calibrated at the factory. A Coriolis meter with advanced sensor diagnostics can be field verified without using a reference standard. This involves four parts:

1) Meter transmitter verification
2) Coriolis sensor verification
3) Temperature verification
4) Meter zero verification (and re-zero if necessary)

Meter transmitter verification should coincide with performing the meter zero verification. This includes confirming the correct configuration and that all diagnostic indicators are in the normal state. A loop check can also be performed to verify the transmitter output at zero and normal flow.

Coriolis sensor verification is a diagnostic feature of some Coriolis meter models. If this feature is available, it checks the stiffness of the meter tubes to detect changes from their condition at the last full calibration. This diagnostic can be run as frequently as desired. If there is no change in tube stiffness, then there will be no change in the calibration factor. If the meter fails the sensor verification three times in a row, then the meter should be sent to a certified calibration lab or the manufacturer.

Coriolis meters monitor a temperature element bonded to the flow tubes and use this to correct for temperature effects on the metal tube stiffness. Use a temperature reference placed in an upstream thermowell near the meter to verify the temperature indicated by the Coriolis meter. The temperature measured at the thermowell should be within the published accuracy of the Coriolis meter temperature sensor and the reference sensor.

Zero verification is used to evaluate the stored factory zero and to determine whether a field zero calibration can improve measurement accuracy. Factory zero is more accurate than field zero. Do not re-zero unless it is required by site procedure (or regulation) or the stored factory zero fails the zero verification procedure.

The zero verification procedure is as follows:

1) Allow the flowmeter electronics to warm up for 20 minutes after applying power.
2) Run process fluid through the meter until the meter's internal temperature sensor reaches the process fluid temperature.

3) Stop the flow through the meter. Close the downstream isolation valve (and upstream valve if available).
4) Verify that the sensor is blocked in, the flow is stopped, and the sensor is full of fluid.
5) Start the zero verification procedure in the transmitter.
6) If it fails the first time, verify that the flow is stopped, the sensor is blocked in and full of fluid, the fluid is not flashing (if liquid) or condensing (if gas), and no particles settle out. This may also indicate that the meter tubes have become coated or fouled internally and need to be cleaned. If possible, clean the meter and try again.
7) If it fails again, follow the manufacturer's procedures to perform a zero calibration.

Review Questions

8.1 Why is in situ calibration often preferred over bench calibration for a flowmeter?
a) It verifies meter accuracy under actual process conditions with the actual process fluid.
b) Field reference standards are cheaper and easier to use.
c) It can be performed without a formal procedure.
d) It eliminates the need to involve maintenance personnel.

8.2 After proving a liquid meter and calculating a meter factor, why might a meter be adjusted or replaced instead of just applying the new meter factor?
a) Meter factors can only be used once.
b) Adjusting the meter is easier.
c) Meter factors are not allowed in regulated applications.
d) A significant change in meter factor from the previous proving may indicate wear, damage, or malfunction.

8.3 To calibrate an orifice meter in gas service, the typical method would be to:
a) Inspect the primary devices (orifice plate and holder) and calibrate the secondary devices (DP, pressure, and temperature sensors).
b) Use meter diagnostics.
c) Flow into a proving tank.

8.4 What is the main purpose of proving a liquid flowmeter?
a) To verify the pressure rating of the meter under maximum flow
b) To determine a correction factor (meter factor) that adjusts the meter's output to match a known reference value
c) To clean the internals of the meter using calibration fluid
d) To verify that the meter's tag number matches the drawings

8.5 How is meter proving accomplished for liquid meters?
a) By comparing the flowmeter's indicated volume to a known reference volume at actual flowing conditions
b) By disconnecting the meter and testing it on a bench with water
c) By averaging the readings from multiple flowmeters in the same line

References

Note: The dates and editions for standards and other references in this chapter are current as of publication. The reader is encouraged to check for later revisions or editions.

1 American National Standards Institute/International Society of Automation (ANSI/ISA), ANSI/ISA 51-1-1979 (R1993), *Process Instrumentation Terminology*. Research Triangle Park, NC: ISA, 1993.

2 American Petroleum Institute (API), *Manual of Petroleum Measurement Standards,* Chapter 22 – "Testing Protocol Section 2 Differential Pressure Flow Measurement Devices," 2nd ed. Washington, DC: API, April 2017, reaffirmed February 2022.

3 American Petroleum Institute (API), *Manual of Petroleum Measurement Standards,* Chapter 14.3.1 – "Orifice Metering of Natural Gas and Other Related Hydrocarbon Fluids – Concentric, Square-edged Orifice Meters – Part 1: General Equations and Uncertainty Guidelines," (ANSI/API MPMS 14.3.1) (AGA Report No. 3, Part 1), 4th ed. September 2012 with July 2013 errata, reaffirmed July 2022. Washington, DC: API.

4 Meter Engineers, (n.d.), *Volumetric Tank Provers: Meter Proving and Calibration Equipment*. https://meterengineers.com/meter-provers/volumetric-tank-meter-provers/. Accessed March 12, 2025.

5 American Petroleum Institute (API), Manual of Petroleum Measurement Standards, Chapter 4.5 – "*Master Meter Provers*," 4th ed. Washington, DC: API, 2016.

6 American Gas Association (AGA), 2003, Report 11 – "Measurement of Natural Gas by Coriolis Meter." Washington, DC, AGA.

9

Final Control Devices Calibration

9.1 What Are Final Control Devices?

The temperature, pressure, level, and flow instruments covered in the previous chapters all sense a process parameter and produce a signal for indication or input to a controller. To control a process parameter, the controller output must be converted to a signal that may be used to drive a final control element. A final control element is any device or element that is used to adjust the process.

Table 9.1 lists some examples of final control elements. This chapter provides examples of control valves and the devices that process the signal supplied to the control valve.

9.2 Flow Control Loop with Control Valve Final Control Element

In Figure 9.1, the controller output sends an electronic signal to the current-to-pressure (I/P) transducer, which sends a pneumatic signal to the control valve. The control valve position changes in response to the signal to adjust the flow to the set point. As the flow changes, it is sensed by the flow transmitter. The valve position is maintained when the flow sensed is equal to the set point. Any time there is a disturbance to the system or a change in the set point, the flow control loop automatically responds to achieve the programmed set point. A block diagram of this concept is illustrated in Figure 9.2.

The final control element can be proportional control, as described, or on-off control. For on-off control, a controller output relay changes the state of the relay contact that completes the circuit for a solenoid valve to energize. The solenoid valve opens to allow air to open (or close) a control valve.

The first component in the final control subsystem is the signal conditioner. The signal conditioner amplifies and, if necessary, converts the signal for compatibility with the actuator. Typical devices used as signal conditioners include I/P transducers, current-to-voltage (I/E) transducers, amplifiers (electronic or pneumatic), relays, digital-to-analog (D/A) converters, and analog-to-digital (A/D) converters. The most common signal conditioner in a proportional control loop is an I/P transducer.

A typical I/P transducer (Figure 9.3) is a force balance device in which a coil is suspended in the field of a magnet. Current flowing through the coil generates axial movement of the coil that causes the beam to move. The beam controls the backpressure against the nozzle by controlling the airflow through the nozzle. This back pressure acts as a pilot pressure to control the outlet pressure. The zero adjustment causes the beam to move relative to the nozzle. The span adjustment is a potentiometer that limits the current through the coil. The I/P transducer must be supplied with instrument air within the range specified by the manufacturer, usually at least 20 psig. The typical

Table 9.1 Examples of final control elements.

Final Control Element	Purpose	Examples
Control valves	Regulate fluid flow, pressure, or temperature	Globe valves, ball valves, butterfly valves
Electric actuators	Convert electrical signals into mechanical motion	Motorized valves, solenoid actuators
Pneumatic actuators	Use compressed air to move control devices	Pneumatic diaphragm actuators, cylinder actuators
Hydraulic actuators	Use hydraulic fluid to move control devices	Hydraulic ram actuators, motor actuators
VFDs	Control motor speed for flow/pressure regulation	VFDs for pumps, VFDs for fans
Heaters	Adjust process temperature	Electric heating elements, steam jackets
Dampers	Control airflow in HVAC or combustion systems	Louvers, guillotine dampers
Pumps	Control liquid flow rate	Centrifugal pumps, peristaltic pumps
Compressors	Regulate gas flow or pressure	Centrifugal compressors, reciprocating compressors
Relays or contactors	Switch electrical circuits	Electromagnetic relays, solid-state relays
Positioners	Adjust valve position for precise control	Electro-pneumatic positioners, digital valve controllers
Solid handling	Control and conveyance of solids	Conveyors, auger feeds, and hopper gates

HVAC, heating, ventilation, and air-conditioning; VFD, variable frequency drive.

I/P transducer is calibrated for a 4–20 mA input = 3–15 psig output. Most I/P transducers can be configured for direct action (output pressure increases as input signal increases) or reverse action (output pressure decreases as input signal increases).

The next component in the final control subsystem is the actuator. The actuator receives the conditioned signal and changes it to some form of mechanical energy or motion. Typical devices used as actuators include solenoids, pneumatic valve positioners, alternating current (AC) and direct current (DC) motors, stepper motors, hydraulic motors, and hydraulic pistons. Many control valves include a pneumatic valve positioner.

A valve positioner (Figure 9.4) is used to increase or decrease the air pressure (from the I/P) operating the control valve actuator. Positioners are generally mounted to the control valve actuator and connected mechanically to the valve stem for position indication. A positioner is a type of air relay that acts to overcome hysteresis, packing box friction, and the effects of pressure drop across the valve. It ensures the exact positioning of the valve stem and provides finer control.

There are many types of positioners. There are mechanical positioners and smart electronic positioners. The basic principles of operation are similar for all types. The instrument pressure (e.g., from an I/P) acts on the input module, which controls the flapper-nozzle system of the relay. Supply pressure is applied to the relay, and the output pressure of the relay is supplied to the control valve actuator. Most positioners can be configured for direct or reverse action. For a

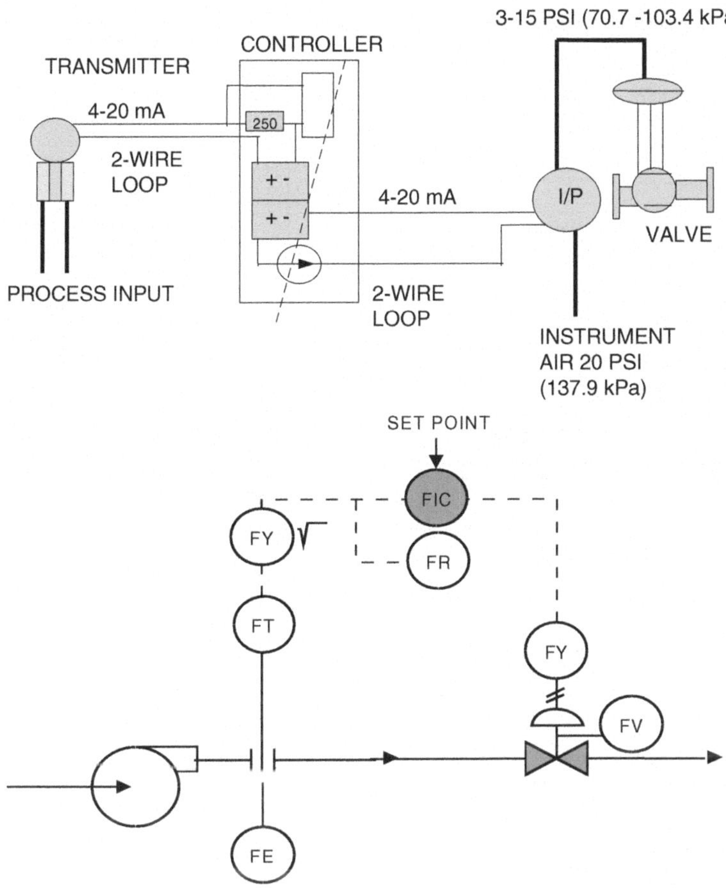

Figure 9.1 Typical control loop components. FE, flow element; FIC, flow indicating controller; FR, flow recorder; FT, flow transmitter; FV, flow valve; FY, flow signal relay device; I/P, current-to-pressure.

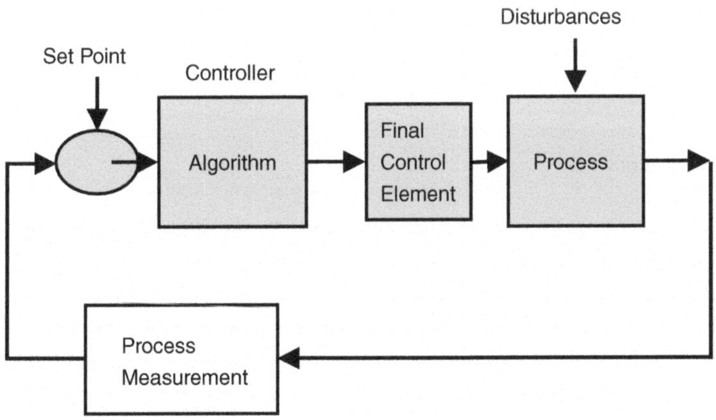

Figure 9.2 Control loop.

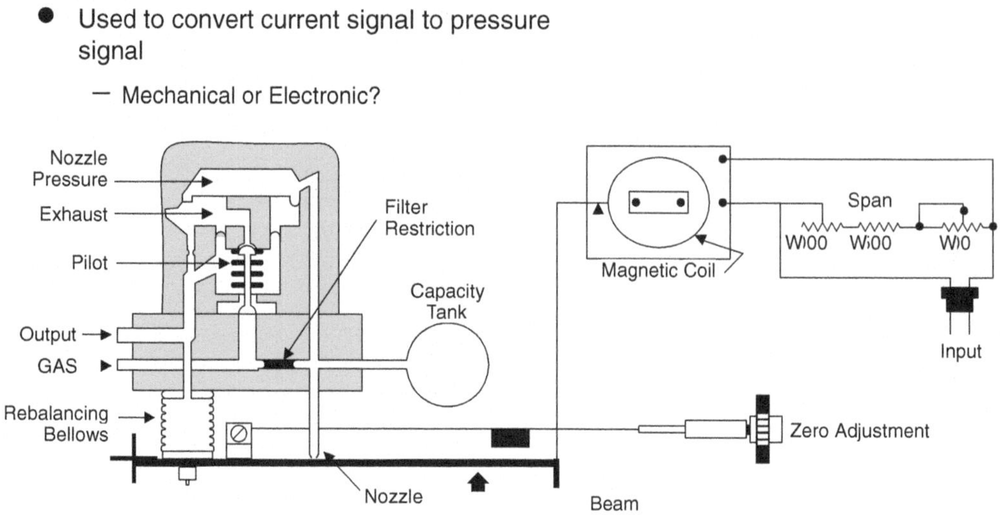

Figure 9.3 Current-to-pneumatic (I/P) transducer.

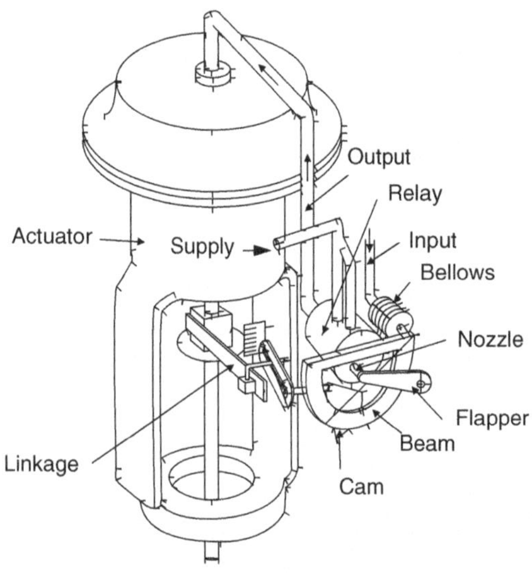

Figure 9.4 Motion-balance mechanical positioner.

direct-acting positioner, increasing the instrument pressure causes the input module to pivot the beam. The beam pivots the flapper and restricts the nozzle. The nozzle pressure increases and causes the relay assembly to increase the output pressure to the actuator. With a direct-acting actuator, the increased pressure moves the actuator stem downward. The positioner is mechanically connected to the stem of the valve. Stem movement is fed back to the beam by a feedback lever and range spring, which causes the flapper to pivot slightly away from the nozzle to prevent further increase in relay output pressure. Some smart positioners accept a milliamp input and include an integral I/P transducer.

The last component in this example is the final control element: a control valve (Figure 9.5). There are many different types, sizes, and applications for control valves. Selecting the correct control

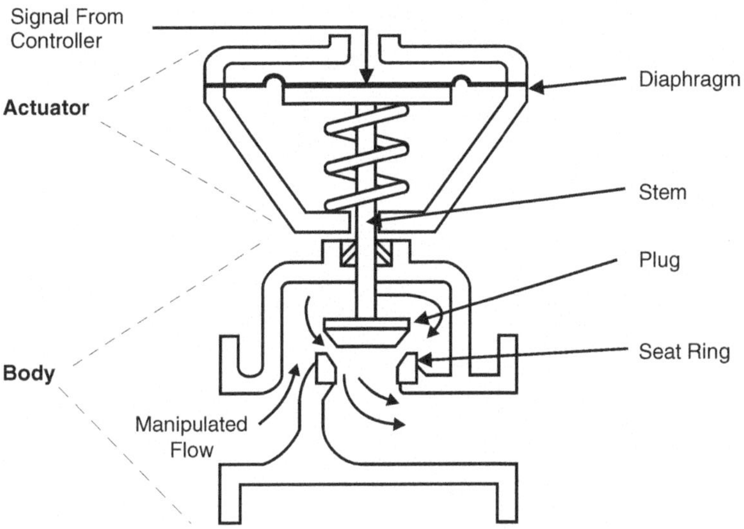

Figure 9.5 Plug-type control valve with diaphragm actuator.

valve for a specific application is crucial to proper system performance. Undersizing and oversizing are common problems. Valve selection is not within the scope of this book; however, ISA's book *Control Valve Primer: A User's Guide* by industry expert Hans Baumann [1] is an excellent resource. Many control valve components must be selected for the application. It is a specialty that is best left to specialists.

The pneumatic signal from the positioner (or I/P transducer if a positioner is not used) is applied directly to the actuator diaphragm. For this control valve, the air is supplied above the diaphragm and pushes against spring pressure to close the valve. The valve is fully closed when the plug is fully seated against the seat ring. Notice that as air pressure is decreased, the spring pressure causes the diaphragm, stem, and plug to move upward, opening the valve. This means that a loss of pressure would cause the valve to open. Therefore, this is known as a fail-open valve. Different configurations of air inlet, spring location, and valve seat arrangement result in different fail positions and determine whether the valve is direct- or reverse-acting. For example, if the plug was below the seat ring (reverse-seated) in the valve pictured in Figure 9.5, the valve would open with increased air pressure and fail closed on loss of air pressure.

All components in the final control subsystem must be configured correctly for the system to work properly. The fail-safe positions must be correct for the application, and the action must produce the desired results. These configurations must be properly documented and used during calibration, loop checks, or troubleshooting.

9.3 Examples

It is possible to do a loop calibration of the control output by putting the control loop in manual, forcing the control system output to a value, and determining whether the valve position is as expected. If the loop calibration fails, these individual device calibrations can be used for troubleshooting.

These examples demonstrate how to calibrate the individual components of a final control system (I/P + positioner + control valve). These are device calibrations.

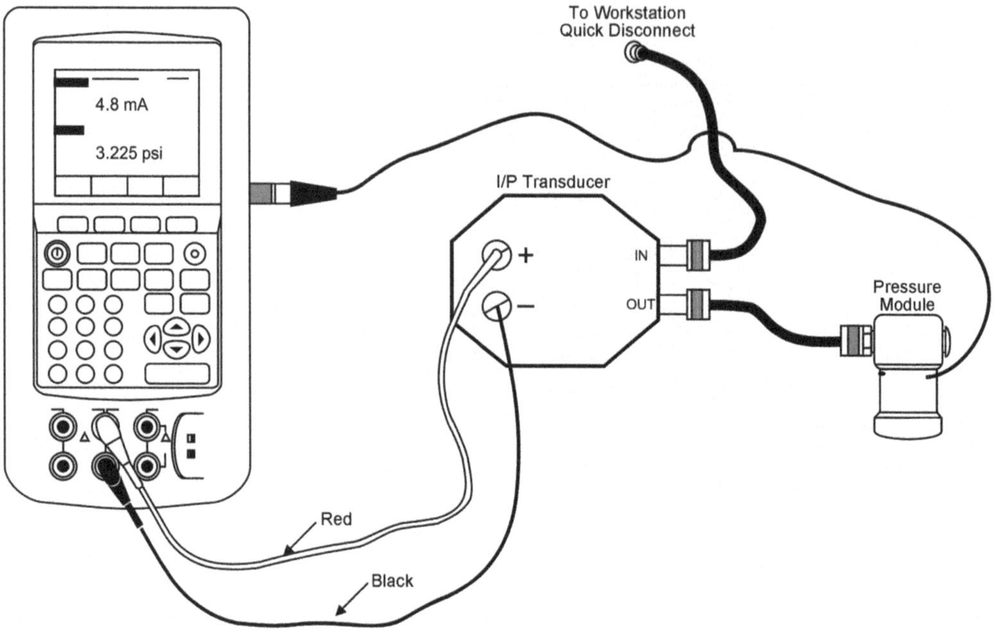

Figure 9.6 Current-to-pressure (I/P) transducer calibration.

Many organizations do not require periodically calibrating I/P transducers, positioners, or control valves. If a control loop is not performing as expected, the output loop and the final control element components should be calibrated. These components are the most common reason for control loops not performing as expected. The mechanical nature of these devices makes them subject to wear, sticking, hysteresis, and so on.

9.3.1 Example: Calibrating an I/P Transducer

Figure 9.6 illustrates the setup for a bench calibration of an I/P transducer. The air supply must be connected to the input according to the manufacturer's specification (typically 20–100 psig). The pressure standard is connected to the air outlet, and a milliamp simulator is connected to the current input. It is important to orient the I/P transducer in the same position it will be in when it is installed in the field. A change in orientation will introduce errors in most I/P transducers.

If the calibration is performed in the field, the existing supply air will be used. It is convenient to tee into the air outlet so the control valve position can be checked at the same time. Of course, you need to ensure the system is in a safe condition before you open and close the valve.

Once the setup is established, apply the milliamp inputs for each desired test point as shown in Table 9.2. Record the corresponding outlet pressure at each test point.

Tech Tip

Some facilities adjust the 0% test point so a slightly higher mA input results in the 0% output. For example, 4.10 mA may result in a 3.0 psig output. This ensures that the valve is closed with a controller output of 4.0 mA.

Table 9.2 I/P calibration test points.

	Applied Input	Expected Output	Actual Output (As-Found)
0%	4.0 mA	3.0 psig	
25%	8.0 mA	6.0 psig	
50%	12.0 mA	9.0 psig	
75%	16.0 mA	12.0 psig	
100%	20.0 mA	15.0 psig	

I/P, current-to-pressure.

Table 9.3 Valve positioner calibration test points.

	Applied Input	Expected Valve Position	Actual Position (As-Found)
0%	3.0 psig	0%	
25%	6.0 psig	25%	
50%	9.0 psig	50%	
75%	12.0 psig	75%	
100%	15.0 psig	100%	

Once the as-found readings are obtained, evaluate the results against the required specification. If necessary, perform zero and span adjustments until no further adjustment is required. Then repeat all test points to record as-left readings.

9.3.2 Example: Calibrating a Valve Positioner

Valve positioner calibration can be performed at the same time as the I/P is calibrated in a loop. Simply tee in the pressure module at the I/P outlet as mentioned in the I/P calibration discussion. Record the valve position at each test point.

If the valve positioner is calibrated separately, connect an input test pressure regulator or hand pump and monitor the input pressure applied with a pressure standard. If supply air is not provided, connect the required supply air to the positioner. Apply the pressure for the desired test points and record the valve positions as shown in Table 9.3.

The valve position indicator on the stem is usually indicated in 5% or 10% increments. In other cases, a valve position detector provides a remote indication to a distributed control system (DCS). In this case, ensure both indicators are working properly. Record the value on the DCS display and the valve stem indicator to check the position feedback.

For guidance for testing positioners, refer to ANSI/ISA 75.13.01-2013, *Method of Evaluating the Performance of Positioners with Analog Input Signals and Pneumatic Output* [2].

9.3.3 Example: Calibrating a Control Valve

As in positioner calibration, a pressure signal is applied to the actuator, and the resulting valve position is recorded. This can be performed with the positioner calibration, if applicable. It can also be performed in conjunction with I/P calibration as described previously. Just remember to ensure the system is in a safe condition if performing the calibration in the field. Also, remember that you should know the correct action, direct or reverse, and fail position before starting.

For guidance on methods for testing positioners and control valve performance, refer to ISA standard ANSI/ISA-75.25.01-2000, *Test Procedure for Control Valve Response Measurement for Step Inputs* [3].

Review Questions

9.1 Refer to the piping and instrumentation diagram (P&ID) in Appendix A. What are the final control elements in temperature loop T-300?

9.2 What is the main objective of calibrating a final control element, such as a control valve?
- a) Ensure the valve is fully closed during shutdown
- b) Verify that the valve position accurately responds to the control signal over its full range of operation
- c) Reduce air consumption in pneumatic actuators
- d) Confirm that the automatic controller is working correctly

9.3 What is the role of a final control element in a control system?
- a) Measure process variables like pressure and temperature
- b) Calculate set points for automatic controllers
- c) Convert control signals into physical action that directly affects the process
- d) Transmit data to the control room for monitoring

9.4 Summarize the basic steps for performing a loop calibration on a control valve.

References

Note: The dates and editions for standards and other references in this chapter are current as of publication. The reader is encouraged to check for later revisions or editions.

1 Baumann, Hans D., *Control Valve Primer: A User's Guide*, 5th ed. Durham, NC: ISA, 2023.

2 American National Standards Institute/International Society of Automation (ANSI/ISA). ANSI/ISA 75.13.01-2013, *Method of Evaluating the Performance of Positioners with Analog Input Signals and Pneumatic Output*. Research Triangle Park, NC: ISA, 2013.

3 American National Standards Institute/International Society of Automation (ANSI/ISA), ANSI/ISA-75.25.01-2000, *Test Procedure for Control Valve Response Measurement for Step Inputs*. Durham, NC: ISA, 2000.

10

Process Analytical Instrument Calibration

10.1 Analytical Applications

There are many specialized analytical instruments used to monitor and control processes. These instruments are less common but no less important than temperature, pressure, level, flow, and final control instruments.

10.2 Analytical Units of Measure

There are too many analytical applications to list all the units of measure here. A few measurement units related to fractional analysis are listed in Table 10.1.

Table 10.1 Analytical units of measure for fractional analysis.

SI Standard Unit of Measure	
N/A	
Other Common Units of Measure	
parts per million (10,000 ppm = 1%)	ppm
parts per billion (10,000,000 ppb = 1%)	ppb
percentage by volume	vol%
percentage by weight	wt%
percentage by mole	mol%
percentage by mass	mass%

N/A, not applicable.

10.3 Types of Analytical Instruments

Analytical instruments serve to measure a variety of fluid properties for product quality, environmental monitoring, engine control, and other purposes. There are applications in liquids and gases. Table 10.2 lists some common examples of analytical instruments.

Table 10.2 Examples of analytical instruments in process control.

Instrument Type	Principle of Operation	Typical Application
Gas chromatograph (GC)	Separation of components in a gas mixture based on volatility and polarity	Monitoring the composition of natural gas or hydrocarbons
Flame ionization detector (FID) on a GC	Ionizes compounds in a flame and measures the ion current	Measuring hydrocarbons in gas streams
Thermal conductivity detector (TCD) on a GC	Detects heat transfer changes due to gas composition	Measuring hydrogen content in gas mixtures
pH meter	Measures the hydrogen ion concentration using an electrode	Acid or base control in chemical processes
Mass spectrometer	Ionizes chemical species and measures their mass-to-charge ratio	Identifying compounds in pharmaceuticals or chemicals
Conductivity meter	Measures the electrical conductivity of a solution	Monitoring water purity or salt concentration
Spectrophotometer	Measures the absorption of specific wavelengths of light by a sample	Measuring concentrations of solutions (e.g., dyes)
Infrared (IR) analyzer	Measures the absorption of infrared radiation by molecules	Detecting CO_2 in flue gases
Turbidity meter	Measures light scattering caused by suspended particles	Monitoring water quality
Nuclear magnetic resonance (NMR) spectrometer	Explores the magnetic properties of nuclei in a magnetic field	Determining the molecular structure of organic compounds
X-ray fluorescence (XRF) analyzer	Excites a sample with X-rays and measures secondary fluorescence	Analyzing metal content in ores or industrial alloys
Dissolved oxygen meter	Measures oxygen levels in liquids using an electrochemical sensor	Monitoring aeration in wastewater treatment
Refractometer	Measures the refractive index of a liquid or solution	Monitoring sugar concentration in food production
UV-Vis spectrometer	Measures the absorption of ultraviolet or visible light by a sample	Analyzing color and chemical concentration
Oxygen analyzer	Measures oxygen concentration in gas using paramagnetic or electrochemical principles	Combustion control in boilers Pipeline protection from corrosion
Moisture analyzer	Measures the dew point or ppm of water in a gas stream	Corrosion prevention Dehydration system monitoring

CO_2, carbon dioxide; ppm, part per million; Uv-Vis, ultraviolet-visible.

As with other instruments, the purpose of calibration of analytical instruments is to compare the instrument reading against a standard. The standard is usually a fluid or a gas of precisely known composition or properties of interest. For example, a sensor for measuring oxygen in a gas stream would be tested in a gas with a known fraction of oxygen. Each type of instrument in Table 10.2 would require a specific calibration standard fluid or gas.

A two-point check is often used for analytical instruments:

- **Zero check** – Apply a zero-level standard (e.g., pure nitrogen for an oxygen sensor or distilled water for conductivity) to confirm the instrument reads zero or the minimum scale value. The zero can be adjusted if needed.
- **Span check** – Apply a known, certified standard representing the upper end of the measurement range (e.g., 100 ppm oxygen or 10.00 pH buffer solution). Allow the instrument to stabilize and record the reading. Adjust the output or replace the sensor if needed.

Many analytical instruments can be calibrated in situ, which allows for true loop calibration. When possible, loop calibration should be performed as it is with other process instruments. This allows testing the control system reading as well as the instrument reading. Some analytical instruments have automated calibration systems. For example, some gas analyzers have systems that periodically apply the standard gas to the analyzer and record the result. These automated systems can alert technicians when the calibration results are not ideal. These systems will require attention and maintenance (e.g., changing the bottle of reference gas on a gas chromatograph [GC] installation).

This chapter gives examples for pH, conductivity, and resistivity process instruments.

10.4 Examples

10.4.1 Example: pH

Many industrial processes, such as food processing, sewage treatment, water purification, and pharmaceutical production, are sensitive to pH. pH is the measure of the acidity or alkalinity of an aqueous solution. Maintaining the proper pH is essential for living systems.

pH is equal to the negative logarithm of the hydrogen ion concentration (Equation 10.1).

$$\mathrm{pH} = -\log[\mathrm{H}^+] \tag{10.1}$$

Hydrogen ion concentrations change when an acid or base is added to water. An acid increases the concentration of hydrogen ions in a solution by donating one of its hydrogen atoms through dissociation. A base decreases the concentration of hydrogen ions by providing hydroxide ions that combine with free hydrogen ions and removes them from the solution. Water automatically dissociates (ionizes) a few of its molecules into hydrogen and hydroxide ions (Equation 10.2).

$$\mathrm{H_2O} \rightarrow \mathrm{H^+} + \mathrm{OH^-} \tag{10.2}$$

Table 10.3 is an example of H^+ concentrations and interpreting the associated pH value. For example, the H^+ concentration of pure water is 1×10^{-7} mol/L. The log of 1×10^{-7} is −7. Therefore, the negative log is 7, and the pH of pure water is 7. A pH of 7 is considered neutral, neither acidic nor basic. Solutions with a pH value less than 7 are acidic; those with pH values above 7 are basic (sometimes called *alkaline*).

Table 10.3 pH values.

Hydrogen Ion Concentration			Nominal Voltage Output @ 25 °C (mV)
Exponential	**Decimal**	**pH**	
		Acidic	
10^{0}	1.0	0	414.12
10^{-1}	0.1	1	354.96
10^{-2}	0.01	2	295.8
10^{-3}	0.001	3	236.64
10^{-4}	0.0001	4	177.48
10^{-5}	0.00001	5	118.32
10^{-6}	0.000001	6	59.16
		Neutral	
10^{-7}	0.0000001	7	0.0
		Basic	
10^{-8}	0.00000001	8	−59.16
10^{-9}	0.000000001	9	−118.32
10^{-10}	0.0000000001	10	−177.48
10^{-11}	0.00000000001	11	−236.64
10^{-12}	0.000000000001	12	−295.8
10^{-13}	0.0000000000001	13	−354.96
10^{-14}	0.00000000000001	14	−414.12

The instrumentation used to measure pH in a process system consists of a pH meter and electrode (Figure 10.1). The electrode is a combination of a measuring electrode and a reference electrode. Both electrodes are incorporated into a single probe in industrial applications for continuous measurement (Figure 10.2).

The measuring electrode has a thin, fragile glass bulb at its tip that is sensitive to the H^+ ions in the surrounding medium. An electrical potential develops between the inner and outer surfaces of the glass bulb. The magnitude of this potential varies depending on the concentration of hydrogen ions in the solution. The reference electrode works the same way, but it is not exposed to the process. It is in a buffer solution of known pH, so it provides a stable, constant voltage.

The two electrodes are connected to the pH meter. The meter measures the voltage difference between the reference and the measuring electrodes, amplifies the signal, and converts the signal to a pH value.

The output of a pH sensor is in millivolts (mV). The relationship between pH and voltage is linear (Figure 10.3).

The voltage difference in millivolts is ideally 0 mV at a pH of 7.0. At 25 °C, the nominal millivolt signal changes by ±59.16 mV per pH unit. The ideal values are shown in Table 10.3.

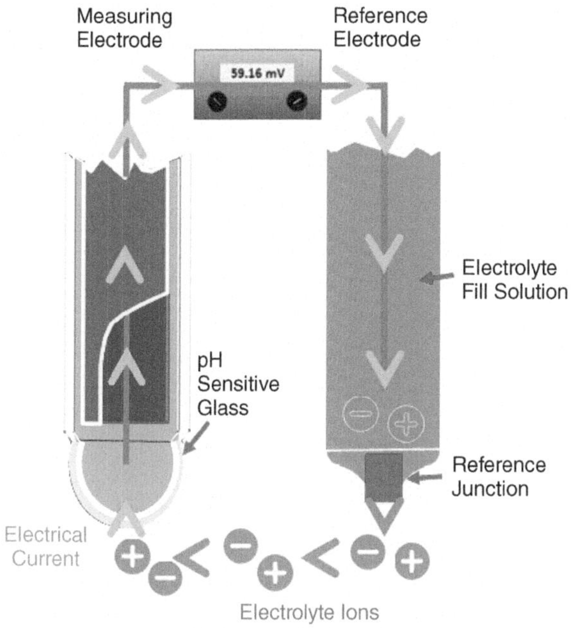

Figure 10.1 pH electrode and meter [1]. "Proven Solutions for pH Measurement," Emerson Automation Experts Blog. Reproduced courtesy of Emerson, www.emerson.com.

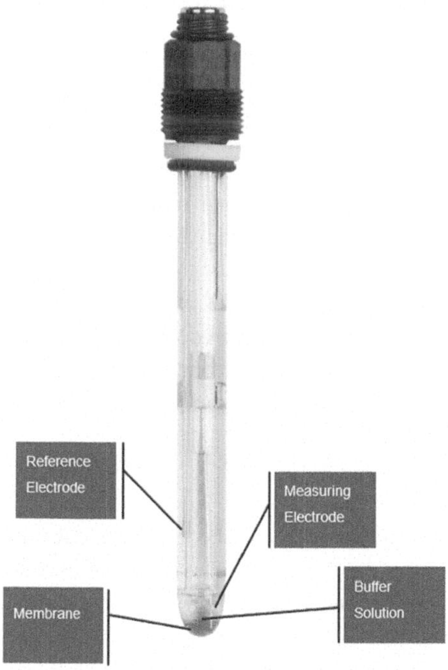

Figure 10.2 pH probe. Isolated On White. Adobe Stock Photos.

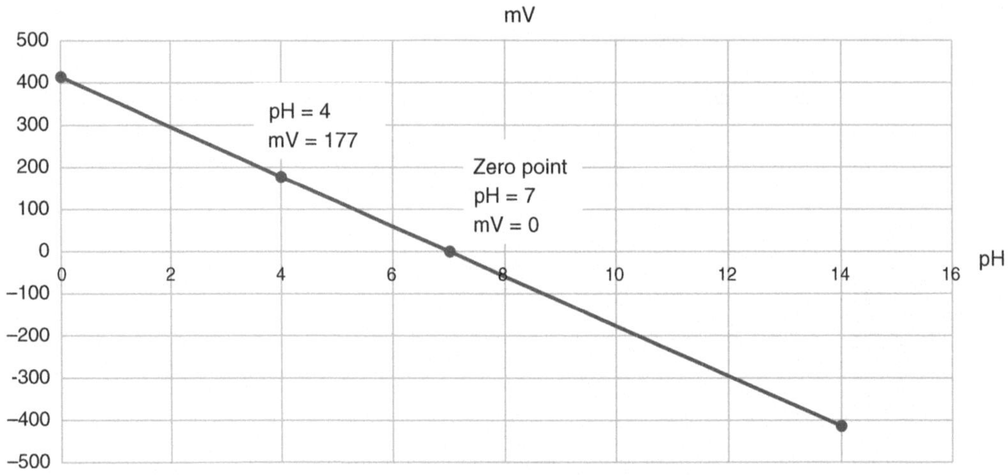

Figure 10.3 Two-point calibration for a pH meter.

The pH meter contains a calibration circuit because the pH electrode is not a perfect device, and performance changes with age. Over time, the slope and offset of the sensor output change. All pH electrodes require calibration from time to time; refer to the manufacturer's recommendations if your organization does not have guidelines.

Because the measurement is linear, a two-point calibration characterizes an electrode with a specific pH meter, as shown in Figure 10.3. When calibrating, the calibration circuit will adjust the millivolt output to read 0 mV while the electrode is immersed in a 7.0 pH buffer standardization solution. This is referred to as the *zero point* or the *standardize* adjustment.

While the electrode is immersed in a 4.0 pH buffer solution, the slope control adjusts the ΔmV output to read +177 mV (or −177 mV). This is referred to as the *slope*, *span*, or *efficiency* adjustment. It is preferable to use a 4.0 pH buffer solution because pH buffer solutions above 7.0 pH are less stable and have a limited life. When left open, these high pH buffers will more readily absorb carbon dioxide (CO_2) from the atmosphere and typically change to a lower pH value.

Temperature is important when measuring pH. The measuring electrode's response to pH is affected by temperature. Also, the pH of the solution that is being measured may increase or decrease as the temperature changes. Most modern pH meters have an automatic temperature compensating (ATC) probe that measures the sample temperature and compensates the pH reading. This ATC probe may be separate or built into the electrode housing.

The chart in Figure 10.4 illustrates the linear ideal pH electrode output at 25 °C and the effect of temperature on the pH electrode output at 0 °C and 100 °C.

pH electrodes must be properly stored when not in use. If not properly stored, the electrode will dry out, which will impair performance or possibly destroy the electrode. The probe should be stored in a 4.0 pH buffer or a 3.8 M solution of potassium chloride (KCl). There are special storage bottles for long-term secure storage to prevent damage and dehydration (Figure 10.5). Do not store in deionized water. Before returning a pH meter to service after storage, the probe should be thoroughly cleaned with deionized water and calibrated.

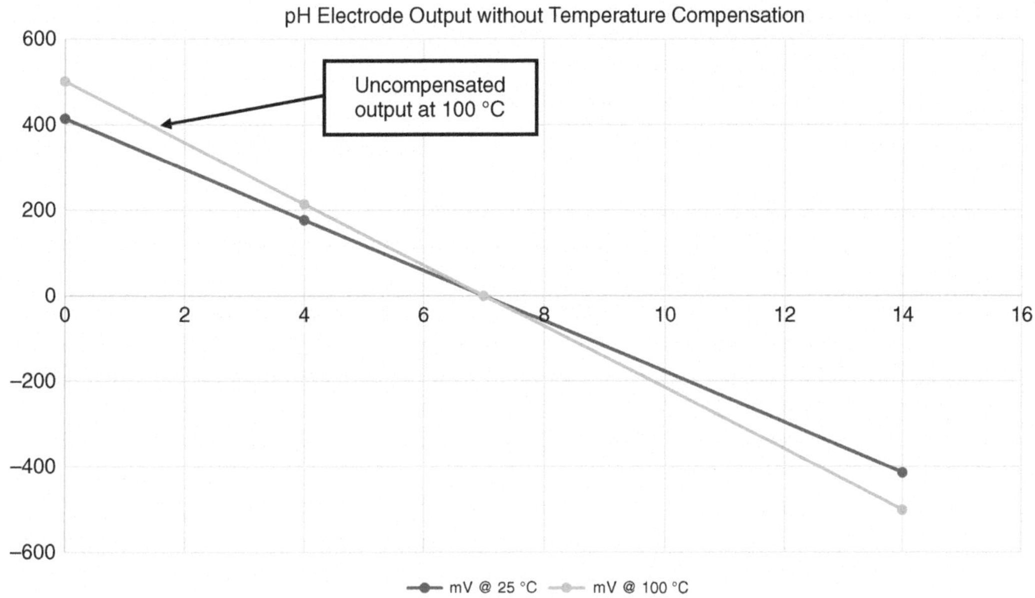

Figure 10.4 Temperature effect on pH probe output.

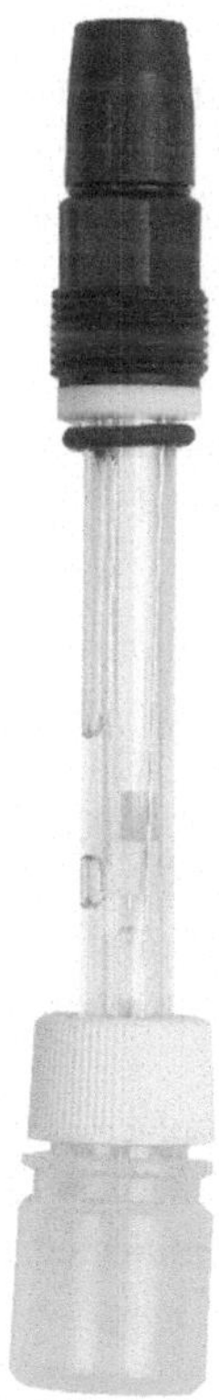

Figure 10.5 pH probe stored in buffer solution. Isolated On White. Adobe Stock Photos.

10.4.2 pH Calibration Procedure

Consult the manufacturer's technical manual for the required calibration procedure, especially for microprocessor-based pH meters with autocalibration features.

1) Rinse the electrode thoroughly with deionized water and gently blot the electrode on a soft tissue.
2) Insert the electrode (and any separate temperature compensation probe) in a 7.00 pH buffer solution. Allow 30 seconds to reach equilibrium or wait for the meter reading to stabilize. Adjust the pH meter for a pH indication of 7.0.
   Note: If the meter does not have automatic temperature compensation, place a calibrated temperature indicator in the same buffer solution with the pH electrode. After 30 seconds, adjust the temperature dial on the pH meter to correspond with the temperature indicated. Then adjust the pH meter for a pH indication of 7.0.
3) Remove the pH electrode from the buffer solution, and rinse and dry it according to the instructions in Step 1.
4) Repeat Step 2, except use a 4.01 pH buffer solution, and adjust for a pH reading of 4.01.
   Note: As mentioned, a 10.0 pH buffer solution may be substituted for the 4.01 buffer solution in Step 4. However, pH buffer solutions above 7.0 are less stable. Therefore, it is preferable to use the 4.01 buffer solution.
5) Repeat Steps 1 through 4 to maximize the precision of the calibration.
6) If desired, check the pH meter indication using a 10.0 pH (or 4.01 pH) buffer solution. Typically, there is no adjustment for this third buffer solution.
   Note: Newer, microprocessor-based pH meters allow the use of any two standard pH buffers that bracket the pH of the samples. Some also allow more than two buffers to determine the slope. Refer to the manufacturer's technical manual for procedures specific to the pH instrument.

10.4.3 Diagnostic Test for pH Electrodes

The asymmetry potential (AP) and slope (efficiency) can be used as a guideline to judge electrode performance. To perform this test, the pH meter must have a millivolt readout.

1) Set the pH meter to read out in millivolts.
2) Connect a shorting plug to the input on the pH meter and adjust the standardize/zero control on the pH meter for a reading of 0.0 mV.
3) Disconnect the shorting plug and connect the pH electrode that will be tested.
4) Rinse the electrode thoroughly with deionized water and gently blot the electrode on a soft tissue.
5) Insert the electrode (and any separate temperature compensation probe) in a 7.00 pH buffer solution. Allow 30 seconds to reach equilibrium. Record the millivolt reading. This is the AP.
   Note: If the meter does not have automatic temperature compensation, place a calibrated temperature indicator in the same buffer solution with the pH electrode. After 30 seconds, adjust the temperature dial on the pH meter to correspond with the temperature indicated. Then, record the millivolt reading to determine the AP.
6) Remove the pH electrode from the buffer solution and repeat Step 4.
7) Repeat Step 5—except use a 4.01 pH buffer solution—and record the millivolt reading.
8) Determine the mathematical difference between the two millivolt readings (from Steps 5 and 7). This is the electrode's span.

9) Divide the electrode's span by the theoretical span of 176.9 mV and multiply the result by 100.

   Example:

   Reading in a 7.00 pH buffer solution: −5.8 mV
   Reading in 4.01 pH buffer solution: +167.5 mV
   AP = −5.8 mV
   Span = 167.5 − (−5.8) = 173.3 mV
   Slope = (173.3 mV/176.9 mV) × 100% = 98%

   Typically, an electrode is replaced when the AP is greater than ±40 mV and/or the slope is <91%.

10.4.4 Measuring Electrical Conductivity and Resistivity

In some situations, it is useful to measure the combined concentration of all ions in an aqueous solution. For example, it can be used to measure the purity of water. The total ionic concentration is determined by measuring the conductivity of the solution. Conductivity is defined as the ability of a material to conduct electrical current. We are used to current flowing through copper wires. Electric current can also flow through ionic solutions (electrolytes). The higher the concentration of ions in a solution, the more conductive the solution.

A simple device to measure conductivity consists of two flat electrodes placed in a sample solution, connected to a battery and an ammeter. The battery generates an electric current that can flow only if the sample solution is conductive and completes the current path. Ohm's law gives the relationship between the voltage and current.

The amount of current flow measured by the conductivity instrument will vary depending on the distance between the electrodes (d) and the area of the electrodes (A). For the same ionic concentration, electrodes spaced farther apart or a smaller electrode area would result in less current measured. The relationship of this distance to the surface area is known as the cell constant (K).

The cell constant is a calibration factor that converts conductance (measured in microsiemens [μS]) to conductivity in a solution (measured in μS/cm). This indicates how well the solution can conduct electricity. The units of cell constant are 1/cm. Normally, the cell constant is measured at the factory. When the sensor is first placed in service, all that is necessary is to enter the cell constant into the analyzer. However, the cell constant can change over time, making recalibration necessary.

A device with a lower constant is more sensitive than one with a higher constant value. Conductivity probes with different cell constants are necessary for the wide range of applications that require conductivity measurement. A purified water system has fewer ions and requires a more sensitive conductivity probe with a K-value of about 0.1. A cell constant of up to 10 would be used for high-conductivity solutions.

Some applications use resistivity instruments. Resistivity instruments are the same as conductivity instruments, except resistivity is simply the inverse of conductivity. Resistivity = 1/conductivity. The units of conductivity are Siemens or mho (℧). In practice, a Siemens is too large, so instruments typically display in microsiemens/micromhos (μS/μ℧) or millisiemens/millimhos (mS/m℧). Resistivity is measured in ohms with typical values in the megohm range: 1 megohm (MΩ) = 1 million ohms.

The conductivities and resistivities of common solutions are listed in Table 10.4.

Like pH, temperature has an effect on conductivity. Conductivity probes include a temperature sensor that measures the temperature of the solution and automatically compensates for the conductivity reading.

Table 10.4 Conductivities of common solutions.

Solution	Conductivity	Resistivity
Ultrapure water	0.055 μS/cm	18.2 MΩ-cm
Distilled water	1 μS/cm	1 MΩ-cm
Deionized water	80 μS/cm	12,500 Ω-cm
0.05% NaCl	1,000 μS/cm	1,000 Ω-cm
Seawater	50,000 μS/cm	20 Ω-cm
30% H_2SO_4	1,000,000 μS/cm	1 Ω-cm

H_2SO_4, sulfuric acid; MΩ, million ohms; NaCl, sodium chloride; μS, microsiemens.

10.4.5 Calibration of Electrical Conductivity

There are two basic ways to calibrate a conductivity meter and sensor:

1) The conductivity system is checked using a standard conductivity solution, if available.
2) The device under test is compared to a calibrated conductivity instrument in the same solution.

Calibrate the temperature sensor against a traceable reference thermometer using a standard solution:

1) Verify that the system is shut down and/or the conductivity cell is isolated from the system pressure. Carefully remove the conductivity probe from the system, ensuring the system pressure is zero.
2) Choose a calibration standard solution near the midpoint of the recommended conductivity range for the sensor. Standard solutions have a shelf life, so make sure yours is not out of date. Use a standard for which the conductivity as a function of temperature is known. A wide range of standard solutions for conductivity calibrations is commercially available. Record the standard solution information, including the expiration date.
3) Turn off automatic temperature compensation in the analyzer if the conductivity of the sample is at the test temperature.
4) Use a calibrated thermometer to measure the temperature of the standard solution. Calculate or look up the actual conductivity in the temperature table provided with the calibration solution. Record the temperature and solution conductivity at that temperature.
5) If the conductivity of the standard is only known at 25 °C, then the temperature of the standard must be adjusted to 25 °C using a water bath.
6) Rinse the beaker and sensor at least twice with the standard.
7) Measure the conductivity using the probe. Record the as-found value and compare it with the standard value.
8) Measure the temperature using the probe. Record the as-found value and compare it with the calibrated thermometer.
9) Adjust the system per the manufacturer's procedure based on the measurement. Record the as-left values of temperature and conductivity.
10) Discard all rinses and the standard after it has been used for calibration. Do not try to recycle the used standard solution.

It might also be necessary to calibrate the built-in temperature sensor using a typical procedure for calibrating temperature.

10.4.6 Calibration Uncertainty for GC

CCST Level III

The overall uncertainty of a GC depends on:

- The precision of the measurement (repeatability)
- The precision of the calibration (repeatability)
- The uncertainty of the calibration gas composition
- Any bias error from the measurement (nonlinearity in a GC)

Notice that the calibration gas composition contributes to the overall uncertainty of the composition analysis. This is important. The chromatograph might be repeatable, but it needs a small target to be accurate.

Here is an example. The goal is to measure the percentage by mole of CO_2 in a carbon capture application. The specification calls for >95 mol% CO_2 in the product stream. The specified measurement tolerance is ±0.25 mol%. The accuracy of the online analysis must be calculated for reporting and must be within the tolerance.

- The manufacturer's stated repeatability for the GC in this example is ±0.1%.
- Use this GC repeatability for the calibration repeatability because the calibration is performed by running calibration gas through the GC.
- The uncertainty of a calibration gas composition is shown on the calibration gas cylinder certificate. For this example, the CO_2 component in the calibration gas has an accuracy of 0.0084% per the certificate. (This is a high-accuracy mix.)
- Assume the bias value is negligible for this example because the component of interest has a very high concentration. Bias might be significant when measuring very low concentrations.

The uncertainty for the CO_2 component is calculated using the root sum squared (RSS) as shown in Equation 10.3.

$$
\begin{aligned}
U_{CO_2} &= \sqrt{R_{\text{meas}}^2 + R_{\text{cal}}^2 + U_{\text{gas}}^2 + \text{bias}^2} \\
U_{CO_2} &= \sqrt{0.1^2 + 0.1^2 + 0.0084^2 + 0^2} \\
U_{CO_2} &= \pm 0.142 \text{ mol\%}
\end{aligned}
\tag{10.3}
$$

This is within the specified tolerance of ±0.25 mol%, but the example demonstrates that the calibration gas accuracy could have a significant impact on the overall measurement uncertainty.

Tech Tip

The calibration gas used in this example is a very high-accuracy ISO Class 1 mixed using a gravimetric method. "Class 1 mixture" refers to calibration gases prepared using the method defined by ISO 6142-1:2015, *Gas Analysis – Preparation of Calibration Gas Mixtures – Part 1: Gravimetric Method for Class I Mixtures*, which is the international standard for high-accuracy gas mixture preparation [2]. The mass of each component added to the mix is carefully measured using traceable mass measurement. The mixtures are then individually verified.

It is important to specify a calibration gas that is mixed with sufficient accuracy that the overall measurement uncertainty is within the specified tolerance of the application. When replacing calibration gas cylinders, be sure to order gas with the correct specification.

Review Questions

10.1 What is the primary objective when calibrating analytical process instruments like pH meters or gas analyzers?
a) Ensure the sensor stays clean during the process operation
b) Verify the instrument's output accurately corresponds to a known reference standard
c) Increase the sample frequency for better process control

10.2 What kind of reference standard is typically used for calibrating analytical instrumentation such as pH meters, conductivity sensors, or gas analyzers?
a) Pressure gauges and temperature probes
b) Electrical signal generators
c) Certified buffer solutions or calibration gases with known chemical compositions
d) Flowmeters and orifice plates
e) Gauge tapes

10.3 Why are zero and span checks, or two-point checks, important when calibrating an analytical instrument?
a) They verify that the instrument responds correctly at the low and high ends of the measurement range.
b) They detect moisture buildup on the sensor.
c) They extend the calibration interval.
d) They determine the loop tolerance.

10.4 Should loop calibration be used on analytical instruments when possible?

References

Note: The dates and editions for standards and other references in this chapter are current as of publication. The reader is encouraged to check for later revisions or editions.

1 Emerson, 2020, "Proven Solutions for pH Measurement," Emerson Automation Experts Blog. https://www.emersonautomationexperts.com/2020/measurement-instrumentation/analytical/proven-solutions-ph-measurement. Accessed August 2, 2025.

2 International Organization for Standardization (ISO), ISO 6142-1:2015, *Gas Analysis – Preparation of Calibration Gas Mixtures – Part 1: Gravimetric Method for Class I Mixtures.* Geneva: Switzerland, ISO, 2015.

11

Fixed Gas Detectors

11.1 What Are Gas Detectors?

Fixed gas detectors are found at many industrial sites. They are devices used to provide warning and activate protection from gas hazards. Typical hazards include:

- **Toxic** – A substance that causes illness or death when inhaled or absorbed by the body in relatively small amounts (e.g., hydrogen sulfide [H_2S]), usually in parts per million
  - Permissible exposure limit (PEL) – The concentration that workers can be exposed to for an 8-hour day, 40-hour week without adverse effects
- **Asphyxiant** – A substance that impairs normal breathing by displacing oxygen (e.g., carbon dioxide [CO_2])
- **Oxygen** – Measured in percentage of volume, alarms occur if there is too much or too little oxygen
  Too little: <19.5% interferes with normal human respiration
  Too much: >25% increases fire hazard
- **Flammable and combustible** – A gas that can ignite and burn (e.g., methane and hydrogen)
  - **Explosive (or flammable) limits** – Desirable gas concentration is outside upper and lower limits – usually in parts per million
  - **Lower explosive** – The minimum concentration of a vapor required to sustain a fire (percentage of material in the air)
  - **Upper explosive** – The maximum concentration of a vapor beyond which a fire cannot be sustained because there is not enough oxygen

Some gases fall into more than one category. For example, hydrogen sulfide is toxic and flammable.

11.2 Gas Detection Terms and Units of Measure

Gas detectors usually give an output that indicates how much of the gas of interest is present. Frequently used units of measure for gas flow are listed in Table 11.1.

11.3 Types of Gas Detectors

Some examples of gas detectors are listed in Table 11.2.

Table 11.1 Units of measurement for gas flow.

SI Standard Unit of Measure	
N/A	
Other Common Units of Measure	
Parts per million (10,000 ppm = 1%)	ppm
Parts per billion (10,000,000 ppb = 1%)	ppb

N/A, not applicable.

Table 11.2 Examples of gas detectors.

Sensor Type	Detects	How It Works
Catalytic bead	• Combustible gas (%LEL)	A catalytic bead increases resistance in proportion to the presence of combustible gas.
Metal oxide semiconductor	• Combustible gas (LEL) • Toxic gas (ppm)	MOS decreases resistance in the presence of gases.
Nondispersive infrared (IR) – point or open path	• Combustible gas (%LEL)	These use two IR beams. The sample beam is at a wavelength absorbed by hydrocarbons, and the reference beam wavelength is outside that range. Detectors compare the ratio of the received signal of each beam.
Tunable laser diode absorption spectroscopy (TDLAS) – open path	• Combustible gas (LEL) • Toxic gas (ppm)	An IR transmitter emits a specific wavelength of IR light that will only be absorbed by the target gas of interest.
Photoacoustic IR	• Combustible gas (LEL) • Toxic gas (%, ppm, ppb)	Gas molecules generate a pressure pulse when exposed to IR light due to heating. The magnitude of the pulse is measured with a microphone.
Electrochemical	• Toxic gas (ppm) • Oxygen (% vol)	These use an electrochemical reaction to generate a current proportional to the gas concentration.

IR, infrared; LEL, lower exposure limit; MOS, metal oxide semiconductor; ppb, parts per billion; ppm, parts per million.

Most sensor types are subject to calibration drift due to the degradation of sensors, exposure to materials that poison the sensors, or harsh operating conditions. Some have a limited shelf life. When an instrument experiences calibration drift, it might still be able to detect a concentration of gas, but it cannot convert it into an accurate reading. Regular calibration with a certified standard gas concentration updates the instrument's reference point so it can give accurate readings. Some instruments have built-in self-test diagnostics that can extend the calibration period.

11.4 Calibration

The gas detector's calibration refers to the instrument's measurement accuracy relative to a known gas concentration. A detector might be adjusted if it fails the calibration test.

Two or more test gases are required.

- **Calibration gas** – A cylinder of gas with a known concentration of the gas of interest that is used to check and set instrument accuracy
- **Zero gas** – A cylinder of gas that is free of the gas of interest that is used to zero an instrument's baseline properly

A bump check verifies that the detector can detect gas, the display is functioning, and the device is measuring accurately. It might also be used to verify that alarms are received. When performing a bump test, the test gas concentration should be high enough to trigger the instrument alarms. Bump checks are usually performed while the instrument is in service.

1) Expose the instrument to a certified calibration gas of known concentration (zero or span gas).
2) Compare the instrument reading with the test gas concentration.
3) Verify that the instrument alarms at the designated alarm set points.
4) Remove the test gas and confirm the instrument returns to zero reading.
5) If the instrument reads and alarms correctly with acceptable accuracy, then it is verified.
6) Record the as-found readings.

The testing frequency depends on the sensor's age, operating conditions, and experience. If it fails a bump test, then perform a full calibration. The unit is sometimes replaced with a spare calibrated unit off the shelf and taken to the shop for calibration. The calibrated unit is put back on the shelf as a spare.

Calibration can include adjusting the instrument's reading to coincide with known concentrations of a certified standard gas.

1) Apply power to the unit to allow the sensor to adapt to the test environment. Usually, sensors must be powered for 1 full hour before calibrating.
2) Check that the instrument reads zero (zero check) where there is no concentration of the target gas.
3) Adjust the zero if needed.
4) Perform a span check by applying gas with the concentration at the full span of the device and adjust the span as needed.
5) Reverify operation with the zero and calibration gases, and then verify the readings and alarms.
6) Record as-left readings.

Review Questions

11.1 Why is it important to regularly calibrate fixed gas detectors in industrial environments?
  a) To reduce the instrument's power consumption
  b) To ensure the detector responds accurately to hazardous gas concentrations and provides reliable alarms for people's safety
  c) To allow operators to disable alarms for maintenance
  d) To adjust the control system's temperature compensation

11.2 Summarize the steps typically involved in calibrating a gas detector.

Appendix A

Piping and Instrumentation Diagram (P&ID)

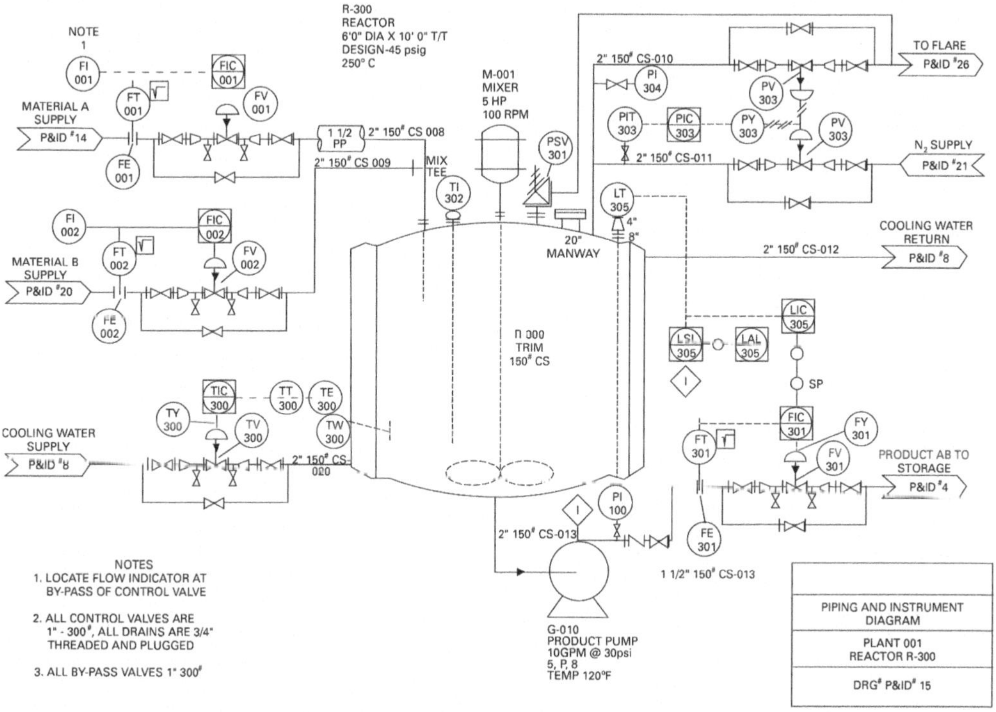

Figure A.1 Piping and instrumentation diagram.

Appendix B

Loop Diagrams

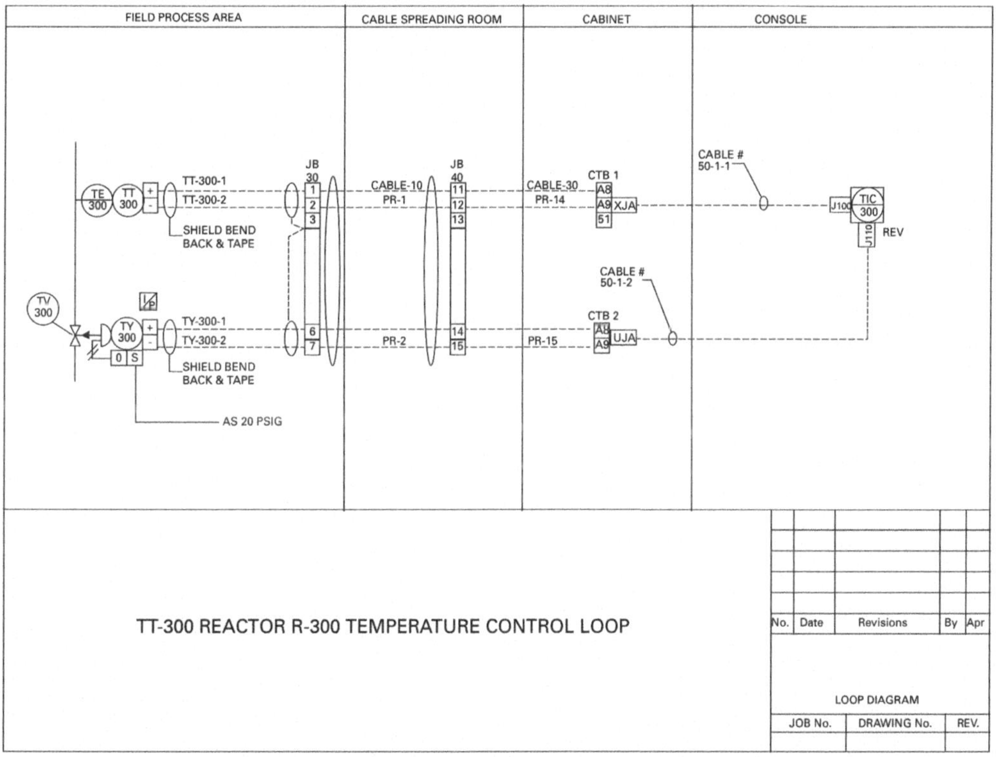

Figure B.1 TT-300 reactor R-300 temperature control loop.

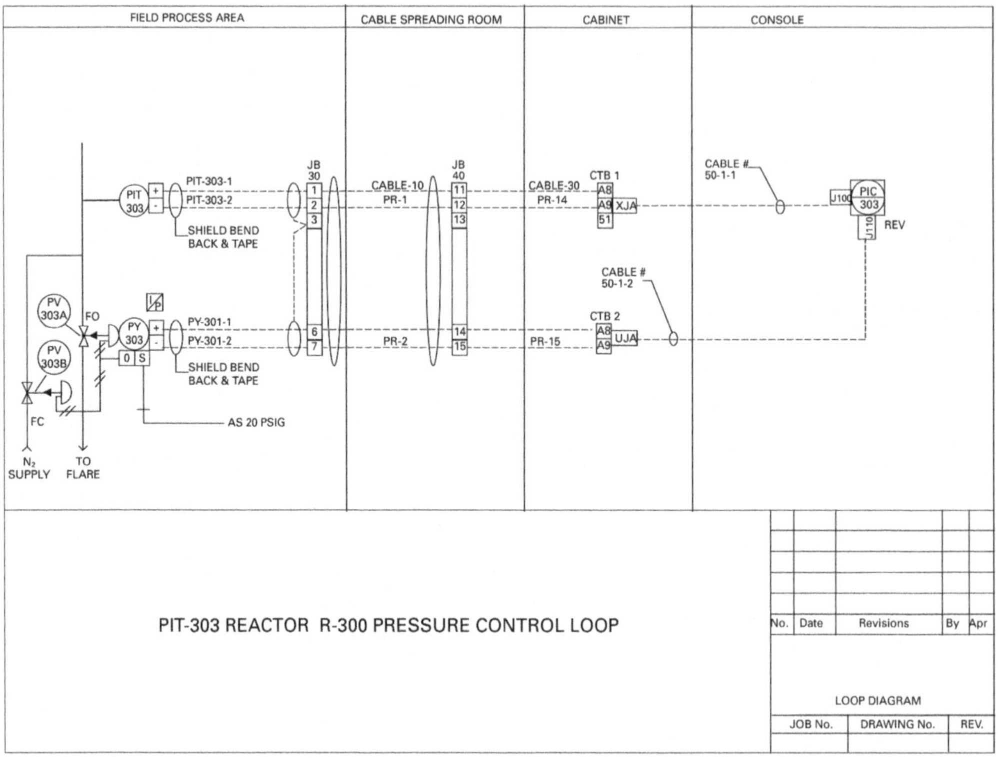

Figure B.2 PIT-303 reactor R-300 pressure control loop.

Appendix C

Instrument Data

This appendix contains details from the manufacturer's data sheets and manuals for the instruments used in the examples.

Temperature sensor: Resistance temperature detector (RTD) – ASTM E1237
Temperature sensor: Thermocouple – ASTM E230
Temperature transmitter – Rosemount 3144P
Pressure gauge – Ashcroft 1279, 0–200 psig
Pressure transmitter – Rosemount 2088
Coriolis meter – Emerson MicroMotion
Valve positioner – Fisher 3582

C.1 RTD Specification

ASTM E1137/E1137M – 08 (2020) Standard Specification for Industrial Platinum Resistance Thermometers

Most manufacturers' specifications will refer to ASTM or International Electrotechnical Commission (IEC) accuracy classes (Figure C.1). This specification [1] lists the ASTM accuracy classes.

Tolerances for platinum resistance thermometers with an average temperature coefficient of resistance 0.385%/ °C between 0 °C and 100 °C:

$$\text{Grade A} = \pm[0.13 + 0.0017|t|]^{\circ}\text{C}$$

$$\text{Grade B} = \pm[0.25 + 0.0042|t|]^{\circ}\text{C}$$

where $|t|$ is the value of temperature in degrees Celsius without regard to sign.

TABLE 1 Classification Tolerances[A,B]

Temperature, *t*, °C	Grade A °C	Grade A Ω	Grade B °C	Grade B Ω
	0.47	0.20	1.1	0.47
	0.30	0.12	0.67	0.27
	0.13	0.05	0.25	0.10
	0.30	0.11	0.67	0.25
	0.47	0.17	1.1	0.40
	0.64	0.23	1.5	0.53
	0.81	0.28	1.9	0.66
	0.98	0.33	2.4	0.78
	1.15	0.37	2.8	0.89
	1.24	0.40	3.0	0.94

[A] The table represents values for 3-wire and 4-wire PRTs. Caution must be exercised with 2-wire PRT's because of possible errors caused by connecting wires.

[B] Tabulated values are based on elements of 100.0 Ω (nominal) at 0 °C.

Figure C.1 Table 1 from ASTM. E1137 [1]. Reproduced with permission from ASTM E230/E230M – 23a, *Standard Specification for Temperature-Electromotive Force (emf) Tables for Standardized Thermocouples*, copyright ASTM International.

C.2 Thermocouple Specification

ASTM E230/E230M – 23a Standard Specification for Temperature-Electromotive Force (emf) Tables for Standardized Thermocouple

Most manufacturers' specifications will refer to ASTM or IEC accuracy classes (Figure C.2). This specification [2] lists the ASTM accuracy classes. ASTM has two accuracy classes for thermocouples: Standard and Special.

Thermo-couple Type	Temperature Range		Tolerances with Reference Junction 0 °C [32 °F]			
			Standard Tolerances		Special Tolerances	
	°C	°F	°C	°F	°C	°F
T	0 to 370	32 to 700	The greater of ±1.0 °C or ±0.75 %	Note 2	The greater of ±0.5 °C or ±0.4 %	Note 2
J	0 to 760	32 to 1400	The greater of ±2.2 °C or ±0.75 %		The greater of ±1.1 °C or ±0.4 %	
*E	0 to 870	32 to 1600	The greater of ±1.7 °C or ±0.5 %		The greater of ±1.0 °C or ±0.4 %	
K or N	0 to 1260	32 to 2300	The greater of ±2.2 °C or ±0.75 %		The greater of ±1.1 °C or ±0.4 %	
R or S	0 to 1480	32 to 2700	The greater of ±1.5 °C or ±0.25 %		The greater of ±0.6 °C or ±0.1 %	
B	600 to 1700	1100 to 3100	±0.5 %		±0.25 %	
C	0 to 2315	32 to 4200	The greater of ±4.4 °C or 1 %		Not applicable	
T[A]	−200 to 0	−328 to 32	The greater of ±1.0 °C or ±1.5 %		[B]	
*E[A]	−200 to 0	−328 to 32	The greater of ±1.7 °C or ±1 %		[B]	
K[A]	−200 to 0	−328 to 32	The greater of ±2.2 °C or ±2 %		[B]	

Figure C.2 Table 1 from ASTM. E230 [2]. Reproduced with permission from ASTM, ASTM E1137/E1137M – 08, *Standard Specification for Industrial Platinum Resistance Thermometers*, copyright ASTM International.

Note 2 to table: At a given temperature that is expressed in degrees Celsius, the tolerance expressed in degrees Fahrenheit is 1.8 times larger than the tolerance expressed in degrees Celsius. Where tolerances are given in percentages, the percentage applies to the temperature being measured when expressed in degrees Celsius. To determine the tolerance in degrees Fahrenheit, multiply the tolerance in degrees Celsius by 9/5.

C.3 Temperature Transmitter Specification

Rosemount 3144P Temperature Transmitter

The information in Figure C.3 is taken from the data sheet for the Rosemount 3144P temperature transmitter [3].

The specifications apply when the transmitter is used with a PT100 RTD input and the 4 mA to 20 mA output.

Input digital accuracy	±0.1 °C (±0.18 °F) for an RTD input
Output digital accuracy	0.02% of span
Input ambient temperature effect (t = ambient temperature)	$0.0015\frac{°C}{°C} \times (t - 20°C)$ $0.0027\frac{°F}{°F} \times (t - 68°F)$ Effect per 1.8 °F (1 °C) change in ambient from calibration temperature (+68 °F/+20 °C): 0.0027 °F (0.0015 °C)
Output ambient temperature effect	D/A Effect: 0.001% of span
Stability	±0.1% of reading or 0.1 °C (0.18 °F), whichever is greater, for 2 years ±0.25% of reading or 0.25 °C, whichever is greater, for 5 years.

Figure C.3 Rosemount 3144P temperature transmitter product specifications. Reprinted courtesy of Emerson, www.emerson.com. D/A, digital-to-analog; RTD, resistance temperature detector.

C.4 Pressure Gauge Specification

Ashcroft 1279 Duragauge Pressure Gauge

The table in Figure C.4 is an excerpt from the Ashcroft 1279 Duragauge Pressure Gauge data sheet [4].

SPECIFICATIONS			
Accuracy:	±0.5% of span (ASME B40.100 Grade 2A)		
Size:	4½″		
Range:	Vacuum, compound to 30,000 psi		
Process Connection Location:	Lower, back, side, top		
Process Connection:	¼ NPT, ½ NPT, 9/16-18 UNF-2B Aminco® (high pressure connection)		
Case Style:	Solid front with pressure relief back		
Window:	Glass (STD.), acrylic, shatter proof glass, non-glare glass (OPT.)		
Movement:	Rotary, adjustable, 400 SS, Teflon™ coated		
Movement Materials:	400 SS, Teflon™ coated pinion gear and segment		
Dial:	Aluminum, white background, black scale		
Pointer:	Micrometer, adjustable, aluminum		
Weather Protection:	Case	IP Rating	Ordering Code
	Dry	Case not sealed, recommended for weather protected environment only	S
	Dry, field fillable	IP66	SH
	Liquid filled	IP66	SL
Dampening Options:	Liquid: glycerin, silicone, Halocarbon®, ***PLUS!***™ Performance		
Mounting:	Stem, surface, flush (STD.), pipe, remote (OPT.)		
Approvals:	CE/UKCA, CRN, NSF-61		

Figure C.4 Ashcroft 1279 Duragauge pressure gauge [4]. Reproduced with permission of Ashcroft Inc.

C.5 Pressure Transmitter Specification

Rosemount 2088 Absolute and Gauge Pressure Transmitter

The specifications in Figure C.5 are taken from the data sheet for the Rosemount 2088 pressure transmitter [5].

Reference accuracy	±0.075% of the calibrated span; includes combined effects of linearity, hysteresis, and repeatability ±0.065% of calibrated span (high-accuracy option –P8) For spans less than 10:1, accuracy = $\pm\left[0.009\left(\frac{URL}{Span}\right)\right]$ percent of span
Ambient temperature effect	Expressed as a total effect per 50°F (28°C) Total effect includes zero and span effects ± (0.15% URL + 0.15% of span)
Stability	Ranges 2–4: ±0.10% of URL for 3years Range 1: ±0.10% of URL for 1year
Vibration effect	Less than ±0.1% of URL when tested per the requirements of IEC 60770-1 field or pipeline with high vibration level (10–60 Hz 0.21 mm displacement peak amplitude/60–2,000 Hz 3g)
Power supply effect	Less than ±0.005% of calibrated span per volt change in voltage at the transmitter terminals

Figure C.5 Rosemount 2088 absolute and gauge pressure transmitter product specifications. *Rosemount 2088 Absolute and Gauge Pressure Transmitter*, product data sheet 00813-0100-4690, Rev QB. Reprinted courtesy of Emerson, www.emerson.com. URL, upper range limit.

C.6 Coriolis Flowmeter

Micro Motion ELITE Series Coriolis Flow and Density Meters

This information consists of two excerpts (Figures C.6 and C.7) from the specification sheet from Emerson [6].

Accuracy and repeatability

Accuracy and repeatability on liquids and slurries

Performance specification	Premium option(1)	Standard option
Mass/volume flow accuracy(2)(3)(4)	±0.05% of rate	±0.10% of rate
Mass/volume flow repeatability	0.025% of rate	0.05% of rate
Density accuracy (2)(4)	±0.2 kg/m³ (±0.0002 g/cm³)	±0.5 kg/m³ (±0.0005 g/cm³)
Density repeatability	0.1 kg/m³ (0.0001 g/cm³)	0.2 kg/m³ (0.0002 g/cm³)

(1) *Not available on all models.*
(2) *For cryogenic applications with process temperatures below −148 °F (−100.0 °C), the liquid mass flow accuracy is ±0.35% of rate, mass flow linearity is ±0.05% of rate, and density accuracy specification does not apply.*
(3) *Stated flow accuracy includes the combined effects of repeatability, linearity, hysteresis, orientation, and other non-linearities.*
(4) *For CMFS007, CMFS010, and CMFS015, the standard volume flow accuracy is ±0.22% of rate and the standard density accuracy is ±2 kg/m³ (±0.002 g/cm³).*

Figure C.6 Specifications for meter accuracy and repeatability with liquids and slurries. *Micro Motion™ ELITE™ Coriolis Flow and Density Meters*, product data sheet PS-00374, Rev AO. Reprinted with permission from Emerson, www.emerson.com.

Accuracy and repeatability on gases

Performance specification	Standard models
Mass flow accuracy(1)	±0.25% of rate
Mass flow repeatability	0.20% of rate
Mass flow linearity	±0.05% of rate up to a 0.2 Mach number
Accuracy with gas calibration linearization(2)	±0.1% of reference after Piecewise Linearization (PWL) adjustment

(1) *Stated flow accuracy includes the combined effects of repeatability, linearity, hysteresis, orientation and other non-linearities.*
(2) *Gas calibration at a third-party gas lab can either be managed by the customer after meter delivery or requested as part of the quote process. PWL and gas calibration specification reflects expected AS-LEFT linearized results relative to the gas lab reference standards. Actual results may vary depending on the uncertainty and stability of the gas laboratory reference standards applied.*

Accuracy and repeatability on temperature

Performance specification	Standard models
Temperature accuracy	±1 °C ±0.5% of reading; BS1904 Class, DIN43760 Class A (±0.15 +0.002 × T °C)
Temperature repeatability	0.2 °C

Figure C.7 Specifications for meter accuracy and repeatability with gases and temperature. *Micro Motion™ ELITE™ Coriolis Flow and Density Meters*, product data sheet PS-00374, Rev AO. Reprinted with permission from Emerson, www.emerson.com.

C.7 Valve Positioner

Fisher 3582 and 3582i Positioners and 582i Electro-Pneumatic Converter

The specifications in Figure C.8, extracted from the data sheet [7], apply only to the Fisher 3582 and 3582i positioners.

Product Bulletin
62.1:3582
November 2019

3582 and 3582i Positioners
D200062X012

Specifications

Note: Specifications for 3582 positioners include 3582A, 3582C, 3582D, 3582G, and 3582NS unless otherwise indicated

Available Configurations

Refer to Type Number Description on page 7

Input Signal

3582
■ 0.2 to 1.0 bar (3 to 15 psig), ■ 0.4 to 2.0 bar (6 to 30 psig), or ■ split range, see table 2.
3582i
4-20 mA DC constant current with 30 VDC maximum compliance voltage, can be split range, see table 2.

Equivalent Circuit for 3582i

120 ohms shunted by three 5.6-volt zener diodes, see figure 2

Output Signal

Type: Pneumatic pressure as required by actuator up to 95 percent of maximum supply
Action: Field-reversible between ■ direct and ■ reverse within the pneumatic valve positioner

Supply Pressure(1)

Recommended: 0.3 bar (5 psi) above actuator requirement
Maximum: 3.4 bar (50 psig) or pressure rating of actuator, whichever is lower

Supply Medium

Air or Natural Gas

The 3582i positioner is not approved for use with Natural Gas as the supply medium

Maximum Input Bellows Pressure Rating(1)

2.4 bar (35 psig)

Maximum Steady-State Air Consumption(2)

3582:
1.4 bar (20 psig) Supply: 0.38 normal m^3/hr (14.0 scfh)
2.0 bar (30 psig) Supply: 0.48 normal m^3/hr (18.0 scfh)
2.4 bar (35 psig) Supply: 0.54 normal m^3/hr (20.0 scfh)

3582i:
1.4 bar (20 psig) Supply: 0.46 normal m^3/hr (17.2 scfh)
2.0 bar (30 psig) Supply: 0.57 normal m^3/hr (21.4 scfh)
2.4 bar (35 psig) Supply: 0.64 normal m^3/hr (23.8 scfh)

Maximum Supply Air Demand(2)

1.4 bar (20 psig) Supply: 4.4 normal m^3/hr (164.5 scfh)
2.0 bar (30 psig) Supply: 6.7 normal m^3/hr (248.5 scfh)
2.4 bar (35 psig) Supply: 7.7 normal m^3/hr (285.5 scfh)

Performance

3582
Independent Linearity: ±1 percent of output signal span
Hysteresis: 0.5 percent of span

3582i
Independent Linearity: ±2 percent of output signal span
Hysteresis: 0.6 percent of span

Electromagnetic Compliance for 582i electro-magnetic converter
Meets EN 61326-1:2013
Immunity—Industrial locations per Table 2 of the EN 61326-1 standard. Performance is shown in table 1 below.
Emissions—Class A
ISM equipment rating: Group 1, Class A

Note: Electromagnetic Compatibility also applies to the 3582i positioner

3582 and 3582i
Open Loop Gain (Output Signal):
■ 100 in the range of 0.2 to 1.0 bar (3 to 15 psig)
■ 55 in the range of 0.4 to 2.0 bar (6 to 30 psig)

Operating Influences

Supply Pressure, For 3582: Valve travel changes less than 1.67 percent per bar (0.25 percent per 2 psi) change in supply pressure
Supply Pressure, For 3582i: Valve travel changes less than 3.62 percent per bar (1.5 percent per 2 psi) change in supply pressure

Figure C.8 Fisher 3582 and 3582i positioner specifications. *Fisher™ 3582 and 3582i Positioners and 582i Electro-Pneumatic Converter*, Product Bulletin 62.1:3582. Reprinted with permission from Emerson, www.emerson.com.

References

Note: The dates and editions for standards and other references in this chapter are current as of publication. The reader is encouraged to check for later revisions or editions.

1 American Society for Testing and Materials (ASTM), 2020, ASTM E1137/E1137M – 08, *Standard Specification for Industrial Platinum Resistance Thermometers.* West Conshohocken, PA: ASTM International.

2 American Society for Testing and Materials (ASTM), n.d., ASTM E230/E230M – 23a, *Standard Specification for Temperature-Electromotive Force (emf) Tables for Standardized Thermocouples.* West Conshohocken, PA: ASTM International.

3 Emerson, 2025, *Rosemount 3144P Temperature Transmitter*, Product Data Sheet 00813-0100-4021, Rev TG, February 2025. Clayton, MS: Emerson.

4 Ashcroft, n.d., *1279 Duragauge Pressure Gauge* data sheet, 279_gauge_ds_RevS_30-07-25. https://www.ashcroft.com/wp-content/uploads/2020/09/datasheet-ash-1279-duragauge-pressure-gauge.pdf. Accessed February 5, 2026.

5 Emerson, 2021, October, *Rosemount 2088 Absolute and Gauge Pressure Transmitter*, Product Data Sheet 00813-0100-4690, Rev QB. https://www.emerson.com/documents/automation/product-data-sheet-rosemount-2088-absolute-gauge-pressure-transmitter-en-73236.pdf. Accessed April 7, 2025.

6 Emerson, 2024, December, *Micro Motion™ ELITE™ Coriolis Flow and Density Meters*, Product Data Sheet PS-00374, Rev AO. https://www.emerson.com/documents/automation/product-data-sheet-micro-motion-elite-coriolis-flow-density-meters-en-66748.pdf. Accessed April 7, 2025.

7 Emerson, 2019, *Fisher™ 3582 and 3582i Positioners and 582i Electro-Pneumatic Converter*, Product Bulletin 62.1:3582. https://www.emerson.com/documents/automation/product-bulletin-fisher-3582-3582i-positioners-582i-electro-pneumatic-converter-en-124122.pdf. Accessed April 7, 2025.

Appendix D

Calibration Procedures

The calibration procedures in this appendix are used in the examples in the chapters. These are organized similar to the example procedure in Appendix A.2 of ISA-RP 105.00.01-2017 [1] and documented on a calibration service record (CSR) form, such as the ISA form shown in Section D.6.

- Temperature Loop Calibration Using a Dry-Block Calibrator
- Pressure Gauge Calibration
- Pressure Loop Calibration
- Level Loop Calibration
- Meter Proving Procedure
- Calibration Service Record Example Form

D.1 Temperature Loop Calibration Using a Dry-Block Calibrator

D.1.1 Description

This loop is used for controlling process temperature, provides temperature indication on the control room operator interface computer (OIC) human-machine interface (HMI), and has alarm points for high and low temperatures. The temperature is critical for product quality.

The Class B PT100 resistance temperature detector (RTD) is connected to a transmitter that is configured to output a 4–20 mA signal representing 0 °C to 200 °C. The current signal is connected to an analog input card on a programmable logic controller (PLC). The OIC HMI display is configured to indicate the temperature from 0.0 °C to 200.0 °C.

The overall loop tolerance is ±1.5 °C.

D.1.2 Preparation

1) Inform the shift supervisor and control operator of the work to be performed. Ensure the loop can be taken out of service during the calibration. (*Note:* This procedure can take up to 2 hours.)
2) Determine what, if any, effect this work will have on the other systems.
3) Review the calibration procedures. Ensure the test uncertainty ratio (TUR) is 4:1 or better for the selected calibration equipment.
4) Conduct a job briefing with all personnel assigned to the work before the job begins.

D.1.3 General

The tasks must be performed by a fully trained control system technician working with either an apprentice control system technician, a calibration technician, or a control operator. The control system technician initiates each step. The calibration technician, apprentice control system technician, or operator will record the data and track the procedure.

The following items are required to perform the work:

- A reference standard, which must have a current certificate traceable to the National Institute of Standards and Technology (NIST), such as a Fluke 9102S handheld, portable dry-block calibrator with a built-in reference sensor. This procedure uses an internal reference sensor.
- Tools to remove the RTD from the thermowell.
- CSR form.
- Task list for calibrating the temperature loop.

D.1.4 Procedure

1) Proceed with the loop calibration task list.
2) If the loop does not meet tolerance, then perform calibration on the loop components.
3) Replace or adjust devices as required to bring the loop into tolerance.
4) Notify the shift supervisor and/or control operator when the calibration process is complete and the device is returned to service.
5) Sign and file the completed CSR.
6) When all tasks are complete, validate that the system has been correctly returned to service.

D.1.5 Temperature Loop Calibration Task List

Setup

1) Locate and verify the identity of the RTD sensor and transmitter. Inspect for visible damage to the thermowell, sensor, wiring, and transmitter.
2) Confirm that the RTD is installed in a thermowell so it can be safely removed from the process pipe, and then remove the RTD. Verify that there are no leaks around the thermowell.
3) Verify that the correct insert is installed in the calibrator to fit the RTD length and diameter.
4) Connect the HART communicator.
5) Insert the RTD into the dry-block (reference standard).
6) Select the units of measure (degrees Celsius or degrees Fahrenheit).

Calibration Steps

1) Adjust the set point to 0% of span. Allow the temperature in the well to stabilize within ±0.1 °F of the set point. This can take up to 20 minutes. Repeat this for 25%, 50%, 75%, and 100% of the span.
2) After the temperature stabilizes at each set point, record the OIC HMI reading in the As-Found Increasing Indication column on the CSR at 0%, 25%, 50%, 75%, and 100% of the range. Verify that the low and high alarms work as expected as the set points are crossed. This procedure is not performed by decreasing set points, only by increasing them.
3) Determine whether the readings are within the required loop tolerance.
4) If the readings are within the required tolerance, then record those readings again under the As-Left column on the CSR and proceed to Post-Calibration.

Troubleshooting

1) Check the RTD, transmitter, and input card using the applicable procedure. Determine which component is out of tolerance.
2) Check the RTD calibration: Disconnect the RTD from the transmitter; connect the RTD to the RTD input of the calibrator; and run the range again, recording the temperature read by the calibrator versus the input temperature.
3) Adjust the transmitter as necessary using the HART communicator and the device manufacturer's instructions to bring the HMI readings within tolerance.
   Note: The method described here is for a loop calibration. If a device calibration is found to be necessary and the device's 4–20 mA output is adjusted, then the reference milliamp measurement must be used to measure the device output. The HART communicator is not a reference standard.
4) Repeat the calibration and troubleshooting steps until the readings are within the required tolerance, and then record those readings in the As-Left Increasing Indication column.

Post-Calibration

1) Disconnect the HART communicator.
2) Remove the RTD from the dry-block and reinstall it in the thermowell.
3) Reconnect all wires that were disconnected for troubleshooting.
4) Verify that there are no leaks around the thermowell.
5) Confirm that the OIC HMI reading reflects current process conditions.

D.2 Pressure Gauge Calibration Using a Hand Pump

D.2.1 Description

The pressure gauge indicates 0 psig to 200 psig. The pressure gauge is installed with a block valve and a bleed valve to allow for calibration in situ.

The required tolerance is the stated accuracy of ±0.5% full scale (±1 psig).

D.2.2 Preparation

1) Inform the shift supervisor and control operator of the work to be performed. Ensure the gauge can be taken out of service during the calibration.
2) Review the calibration procedures.
3) Ensure the test accuracy ratio (TAR) is 4:1 or better for the selected calibration equipment.
4) Verify that the test equipment is within its calibration period.
5) Conduct a job briefing with all personnel assigned to the work before the job begins.

D.2.3 General

The tasks must be performed by a fully trained control system technician working with either an apprentice control system technician, a calibration technician, or a control operator. The control system technician initiates each step. The calibration technician, apprentice control system technician, or operator will record the data and track the procedure.

The following items are required to perform the work:

- Reference pressure gauge calibrated and traceable to a standard: Fluke 700G test gauge
- Pneumatic test pump capable of generating the required pressure range: Fluke 700PTP-1
- Fittings to connect both gauges to the hand pump
- Suitable tubing and connectors
- Wrench or tools for secure connections
- CSR form
- Task list for calibrating the pressure gauge

Safety Precautions

- Ensure all equipment is rated for the pressure range to be used.
- Check for leaks in the system before starting the calibration.
- Wear appropriate personal protective equipment (PPE) if required.

D.2.4 Procedure

1) Proceed with the calibration task list.
2) If the gauge does not meet tolerance, then replace the gauge with a calibrated gauge.
3) When all tasks are complete, validate that the gauge has been correctly returned to service.
4) Notify the shift supervisor and/or control operator when the calibration process has been completed and the device has been returned to service.
5) Sign and file the completed CSR.

D.2.5 Pressure Gauge Calibration Task List

Setup

1) Inspect the pressure gauge, and ensure it is free of visible damage.
2) Close the block valve under the gauge.
3) Open the bleed valve under the gauge to bleed the pressure between the block valve and the gauge.
4) Connect the hand pump to a T-fitting or manifold.
5) Attach the reference pressure gauge to one port of the T-fitting.
6) Attach the other port of the T-fitting to the bleed valve for the gauge to be calibrated.
7) Tighten all connections securely to prevent leaks.

System Leak Test

1) Operate the hand pump to pressurize the system to a low pressure (e.g., 10% of the maximum range).
2) Check for leaks at all connections using a soap solution or by observing the gauges for pressure drop.
3) If leaks are found, then release the pressure, tighten the connections, and repeat the test.

Calibration Steps

1) Test gauge zero adjustment
   - With no pressure applied, ensure both the reference gauge and the test gauge read zero.
   - If the test gauge has an adjustable zero, then align it as needed.

2) Pressure application
   - Gradually increase the pressure using the hand pump in steps of 0%, 20%, 40%, 60%, 80% and 100% of the full scale.
3) Record readings
   - At each pressure step, allow the system to stabilize.
   - "At each test point, the gauge under test shall be read, lightly tapped, and then read again. (The difference in the readings is friction error.)" [2]
   - Record the readings of the reference gauge and the test gauge.
4) Descending check
   - Reduce the pressure in steps 100%, 80%, 60%, 40%, 20%, and 0% of full scale, and record the reference gauge and the test gauge readings again.

Final Verification

1) Calculate the error in engineering units (EU) for each step, ascending and descending.
2) Ensure the test gauge readings match the reference gauge within the acceptable tolerance.
3) Compare the ascending and descending readings of the test gauge to identify any hysteresis.

Adjustment

1) If the test gauge readings deviate from the reference, then adjust the gauge according to the manufacturer's instructions (if it is adjustable).
2) Repeat the calibration steps to verify accuracy after adjustments.

Post-Calibration

1) Release all pressure from the system.
2) Disconnect all components and clean the equipment.
3) Label the calibrated pressure gauge with a calibration sticker, including the date, technician's name, and next due date.
4) Close the bleed valve.
5) Open the block valve.
6) Ensure the gauge is reading the expected pressure.

D.3 Pressure Loop Calibration Using a Multifunction Calibrator

D.3.1 Description

This loop provides pressure indications on the control room HMI and has a high alarm point. The pressure is not used for control; it is used for environmental monitoring.

The pressure transmitter is configured to output a 4–20 mA signal representing 0 psig to 200 psig. The current signal is connected to an analog input card on a PLC. The pressure transmitter is installed with a block valve and a bleed valve to allow for calibration in situ.

The OIC HMI display is configured to indicate the temperature from 0.0 psig to 200.0 psig.

The overall loop tolerance is ±2.0 psig.

D.3.2 Preparation

1) Inform the shift supervisor and control operator of the work to be performed. Ensure the loop can be taken out of service during the calibration.

2) Determine what, if any, effect this work will have on the other loops and systems.
3) Review the calibration procedures.
4) Ensure the TUR is 4:1 or better for the selected calibration equipment. Confirm that the calibration test equipment is within its calibration period.
5) Conduct a job briefing with all personnel assigned to the work before the job begins.

D.3.3 General

The tasks must be performed by a fully trained control system technician working with either an apprentice control system technician, a calibration technician, or a control operator. The control system technician initiates each step. The calibration technician, apprentice control system technician, or operator will record the data and track the procedure.

The following items are required to perform the work:

- Automated pressure calibrator (with a range suitable for the transmitter): Fluke 729 automatic pressure calibrator
- Fittings to connect the pressure source to the bleed valve under the transmitter
- Digital multimeter (if necessary for troubleshooting)
- CSR form (to record calibration results)
- Task list for calibrating the pressure gauge

Safety Precautions

- Ensure all equipment is rated for the pressure range to be used.
- Review the Safety Data Sheet (SDS) for the process fluid (be prepared in case of exposure).
- Check for leaks in the system before starting the calibration.
- Wear appropriate PPE if required.

D.3.4 Procedure

1) Proceed with the calibration task list.
2) If the loop does not meet tolerance, troubleshoot and then adjust or replace equipment as needed to bring the loop into tolerance.
3) When all tasks are complete, validate that the loop has been correctly returned to service.
4) Notify the shift supervisor and/or control operator when the calibration process has been completed, and the loop has been returned to service.
5) Sign and file the completed CSR form.

D.3.5 Pressure Loop Calibration Task List

Setup

1) Verify that the transmitter is accessible and correctly identified by its tag.
2) Inspect the pressure gauge, and ensure it is free of visible damage.
3) De-energize any control actions that may be impacted by the calibration procedure (if applicable).

4) Ensure the process is isolated, and the loop is safe to work on (de-energized or in maintenance/calibration mode).
5) Set up communication with the PLC or HMI to monitor the input values.
6) Close the block valve under the gauge.
7) Open the bleed valve under the gauge to bleed the pressure between the block valve and the gauge.
8) Confirm that the correct pressure measurement module is installed in the calibrator.
9) Connect the automated pressure calibrator to the pressure transmitter using appropriate fittings to ensure no leaks.
10) Configure the pressure calibrator to measure the transmitter's 4–20 mA output (optional).

System Leak Test

1) Using the calibrator, apply a low pressure (e.g., 10% of the maximum range).
2) Check for leaks at all connections using a soap solution or by observing the gauges for pressure drop.
3) If leaks are found, then release the pressure, tighten the connections, and repeat the test.

Calibration Steps

1) Pressure application
   - Set the calibrator to apply pressures at test points across the transmitter's operating range: 0%, 25%, 50%, 75%, and 100% of the transmitter's range or as specified on the CSR form.
2) Record readings
   - At each pressure step, allow the system to stabilize.
   - At each test point, read the OIC HMI reading and the calibrator reading.
   - Record the 4–20 mA output from the transmitter (optional).
3) This procedure only checks increasing pressure. Decreasing pressures are not checked.

Final Verification

1) Calculate the error in EU for each ascending and descending step, and record the errors on the CSR form.
2) Ensure the HMI readings match the reference gauge within the acceptable tolerance.

Adjustment

1) Replace or adjust devices as required to bring the loop into tolerance.
2) Adjust the transmitter (if necessary). Use the zero and span adjustment options on the transmitter (if available) to align the output with the correct milliamp signal for each test point.
3) Repeat the calibration steps to verify accuracy after adjustments.
4) Document the final as-found and as-left readings, and ensure the HMI readings accurately reflect the transmitter output.

Post-Calibration

1) Release all pressure from the system.
2) Disconnect all components and clean the equipment.
3) Label the calibrated pressure transmitter with a calibration sticker, including the date, technician's name, and next due date.

4) Close the bleed valve.
5) Open the block valve.
6) Ensure the transmitter and HMI are reading the expected pressure.

D.4 Level Loop Calibration Using a Gauge Tape

D.4.1 Description

This is an innage calibration procedure for a level loop on a custody transfer tank. This loop is for measurement, surveillance, and accounting. There are no control or alarm functions. The level is displayed on the OIC HMI.

For level measurements used in custody transfer or inventory control service, the accuracy is verified at a randomly chosen single level close to the normal opening and closing gauge readings.

This is for the periodic in-service calibrations, not the initial calibration after commissioning. The initial verification procedure requires verification at two points.

Perform a calibration check monthly. The accuracy should be within 3/16 in.

D.4.2 Preparation

1) Inform the shift supervisor and control operator of the work to be performed. Ensure the loop can be taken out of service during the calibration.
2) Determine what, if any, effect this work will have on the other loops and systems.
3) Review the calibration procedures.
4) Confirm the gauge tape has a current calibration sticker.
5) Conduct a job briefing with all personnel assigned to the work before the job begins.

D.4.3 General

The tasks must be performed by a fully trained control system technician working with either an apprentice control system technician, a calibration technician, or a control operator. The control system technician initiates each step. The calibration technician, apprentice control system technician, or operator will record the data and track the procedure.

The following items are required to perform the work:

- Gauging tape
- CSR form

Safety Precautions

- Review the SDS for the process fluid (be prepared in case of exposure).
- Wear appropriate PPE.

D.4.4 Procedure

1) Proceed with the calibration task list.
2) If the loop does not meet tolerance, troubleshoot and then adjust or replace equipment as needed to bring the loop into tolerance.

3) When all tasks are complete, validate that the loop has been correctly returned to service.
4) Notify the shift supervisor and/or control operator when the calibration process has been completed, and the loop has been returned to service.
5) Sign and file the completed CSR.

D.4.5 Automatic Tank Gauge Loop Calibration Task List

Setup

1) Verify the instrument tag. Inspect for visible damage to instrumentation, connections, and wiring.
2) Set up communication with the PLC or HMI to monitor the input values.
3) The tank level should be close to the normal opening or closing gauge reading (for custody application).

Calibration Steps

1) Record the stable level reading immediately before making the manual reference measurements. Check whether the presence of the gauge on the tank top affects the level reading. If the level reading varies by more than 1 mm (1/16 in.), then investigate the cause before proceeding.
2) Determine the reference manual innage measurement of the tank contents from the same gauging access point (using the same measurement tape and weight) until three consecutive measurements agree within a range of 1 mm (1/16 in.) or five consecutive measurements agree within a range of 4 mm (3/16 in.). (See "Manual Gauging Procedure" in this section.)
   Note: Failing to achieve manual measurements within these tolerances may be due to adverse weather conditions, movement of the liquid surface, or inconsistencies in the measurement technique. The verification procedure should be repeated after corrective action has been taken.
3) Calculate the arithmetic average value of the innage measurement (i.e., the average of the qualified, consecutive measurements that agree within the specified tolerance). Do not round this result.
4) Record the level reading immediately after making the manual reference measurements, and confirm that no change has occurred during the manual gauging. If the level reading has changed, check that there has been no transfer either to or from the tank and that the tank valves are closed. Repeat the procedure.
5) Compare the level reading with the average manual innage measurement. The difference between these two measurements is referred to as the *test difference*.

Manual Gauging Procedure

1) After safely grounding the tape and opening the gauge hatch, apply product paste as appropriate and slowly lower the bob and tape into the tank until the bob is within a short distance of the bottom, as determined by the length of tape unwound from the reel in comparison to the reference gauge height of the tank.
2) Then, with the tape adjacent to the reference gauge point, lower the tape slowly until the tip of the bob just touches the datum plate (or tank bottom if no datum plate exists).
3) Record the tape reading at the reference gauge point, and note any variance from the reference gauge height of the tank. Comparing the observed gauge height tape reading to the tank

reference gauge height indicates whether the gauge bob is suspended in a vertical position while in contact with the datum plate or tank bottom. If the tape is lowered too far, causing the bob to tilt, or if the bob is resting on foreign material in the bottom of the tank, then the gauge reading will be inaccurate.

4) When obtaining innage gauges, be sure the tape is lowered at the same reference gauge point for all readings. Allow sufficient time for the surface of the liquid to settle after the bob breaks the surface before continuing to lower the bob.
5) Withdraw the tape from the tank until the liquid cut is observed.
6) Read the tape scale at the liquid cut and note this reading as the innage gauge.
7) Repeat the procedure to obtain the three acceptable readings.

D.5 Meter Proving Procedure for Crude Oil Custody Meter Using a Portable Displacement Prover

D.5.1 Description

This is a calibration procedure for a flowmeter in a custody transfer service. This loop is for measurement and surveillance. There are no control or alarm functions. The flow total is displayed on the local flow computer.

This procedure is for periodic in-service calibrations and not the initial calibration after commissioning. It ensures the accuracy of crude oil custody transfer measurements, maintaining operational integrity and compliance with industry standards.

A calibration check should be performed monthly.

D.5.2 Preparation

1) Inform the shift supervisor and control operator of the work to be performed. Ensure the loop can be taken out of service during the calibration.
2) Determine what, if any, effect this work will have on the other loops and systems.
3) Verify compliance: Ensure all equipment and procedures comply with relevant industry standards.
4) Inspect the equipment.
   - Check the portable displacement prover for cleanliness, integrity, and calibration status.
   - Verify the custody transfer meter's operating condition and connections.
5) Address the environmental conditions.
   - Record ambient and crude oil temperature.
   - Minimize external influences such as vibration or temperature fluctuations.
6) Secure the setup.
   - Position the prover close to the custody meter to minimize pressure loss.
   - Connect the prover to the meter, ensuring all connections are leak-free.
   - Confirm pressure and temperature transmitters are operational.
7) Stabilize the flow.
   - Operate the custody meter at normal process conditions to stabilize the flow rate.

D.5.3 General

1) Use appropriate PPE.
2) Handle crude oil spills or leaks promptly and according to safety protocols.
3) Perform routine maintenance and calibration checks on the prover and custody meter.

D.5.4 Proving Procedure

1) Establish flow rate
   - Set the flow rate at the custody meter to normal operating levels.
   - Allow sufficient time for stabilization.
2) Record baseline parameters
   - Measure and log the initial flow rate, pressure, and temperature from the custody meter.
3) Initiate prover run
   - Activate the prover's piston or ball to begin displacement.
   - Allow the prover to displace a measured volume of crude oil through the meter.
   - Ensure a smooth, uninterrupted flow.
4) Capture proving data
   - Record the meter's pulse count or indicated volume during the prover's operation.
5) Repeat the test
   - Conduct at least three proving runs under identical conditions to ensure repeatability.
   - Confirm the results are within acceptable tolerance limits (e.g., ±0.05%).
6) Calculate meter factor
   - Correct the prover volume for pressure, temperature, and compressibility effects.
   - Use this equation to calculate the meter factor:

$$\text{meter factor} = \frac{\text{prover volume corrected}}{\text{meter-indicated volume}}$$

D.5.5 Post-Proving

1) Analyze results
   - Evaluate the consistency of the meter factor across runs.
   - Investigate anomalies if results exceed tolerance limits.
2) Adjust the meter
   - Update the custody meter's configuration with the calculated meter factor.
3) Document findings
   - Record all test data, calculations, and observations in a proving report.
   - Include pressure, temperature, density, and flow conditions.
4) Return to service
   - Safely drain and disconnect the prover.
   - Resume normal operations.

D.6 Example CSR Form

Figure D.1 is a CSR form from Annex A.3 of ISA-RP105.00.01-2017 [1] used throughout the examples in this appendix. Figure D.2 is an example of a completed form.

A.3 Calibration service record example form

Plant:	Instrument Tag:
Operating Unit:	IO Point ID:
Location:	
Manufacturer:	Calibration Date:
Model:	
Serial #:	Next Calibration Due:

Calibration Location (in-situ, shop, etc.):

Loop/device required tolerance ____ % of ____

	Input	Control System Indication
LRV		
URV		
Eng Units		

Input %	Applied Input (Standard)	Desired Indication	As Found						As Left					
			Increasing Indication	Error in EU	Error in %	Decreasing Indication	Error in EU	Error in %	Increasing Indication	Error in EU	Error in %	Decreasing Indication	Error in EU	Error in %
0														
25														
50														
75														
100														

Test Equipment

Manufacturer	Model	Serial	Certification No.	Certification Date	Accuracy (>4x loop / device)	Remarks

Special Instructions:

Comments:

Calibrated by:	Date:

Figure D.1 CSR form from ISA-RP105. © International Society of Automation. Used with permission of ISA.

Plant:	Instrument/Loop Tag: **PIT-303 / PIC-303**
Operating Unit:	IO Point ID: **CTB1 A8 - PIC-303**
Location:	
Manufacturer: **Rosemount**	Calibration Date: **Nov 1, 2024**
Model: **2088 G 3 S**	
Serial #: **(record serial number)**	Next Calibration Due: **Nov 1, 2025**

Calibration Location (in-situ, shop, etc.): **In situ, loop calibration**

Loop/device required tolerance **±2 psig**

	Input	Control System Indication
LRV	**0**	**0**
URV	**200**	**200**
Eng Units	**psig**	**psig**

Input %	Applied Input (Standard)	Desired Indication	As Found						As Left					
			Increasing Indication	Error in EU	Error in %	Decreasing Indication	Error in EU	Error in %	Increasing Indication	Error in EU	Error in %	Decreasing Indication	Error in EU	Error in %
0														
25														
50														
75														
100														

Test Equipment

Manufacturer	Model	Serial	Certification No.	Certification Date	Accuracy (>4x loop / device)	Remarks
Fluke	**729**	**xxx**	**xxx**	**1-Oct-24**	**±0.02% FS = 0.06 psig**	**use 300 psi module**

Special Instructions: **Calibration procedure SOP-CAL-01**

Comments:

Calibrated by:	Date:

Figure D.2 Completed CSR for the pressure loop example. © International Society of Automation. Used with permission of ISA.

References

Note: The dates and editions for standards and other references in this chapter are current as of publication. The reader is encouraged to check for later revisions or editions.

1. International Society of Automation (ISA), ISA-RP105.00.01-2017, *Management of a Calibration Program for Industrial Automation and Control Systems*. Research Triangle Park, NC: ISA, 2017.
2. American Society of Mechanical Engineers (ASME), ASME B40.100-2022, *Pressure Gauges and Gauge Attachments*. New York, NY: ASME, 2022 Clause 2-5.2.4.1 (b) (2).

Appendix E

Test Equipment

This appendix contains details from the manufacturer's data sheets and manuals for the test equipment used in the examples.

- Temperature calibrator: Beamex MC6 portable dry-well
- Temperature calibrator: Fluke 9102S handheld dry-well
- Pneumatic hand pump: Fluke 700PTP-1 with 700GA27 test gauge – example – 6.5.4.1
- Pressure calibrator, full function: Fluke 729 – example 6.5.4.2

E.1 Temperature Dry-Block Calibrator

Beamex MC6-T Dry-Block Field Calibrator Data Sheet

TEMPERATURE SPECIFICATIONS

FEATURE	MC6-T150	MC6-T660
Temperature range at 23 °C (73 °F)	-30 ... 150 °C (-22 ... 302 °F)	50 ... 660 °C (122 ... 1220 °F)
Display uncertainty with internal reference [1]	±0.15 °C	±0.2 °C at 50 °C ±0.3 °C at 420 °C ±0.5 °C at 660 °C
Stability [2]	±0.01 °C	±0.02 °C at 50 °C ±0.03 °C at 420 °C ±0.04 °C at 660 °C
Axial uniformity at 40 mm	±0.05 °C	±0.05 °C at 50 °C ±0.25 °C at 420 °C ±0.40 °C at 660 °C
Axial uniformity at 60 mm	±0.07 °C	±0.10 °C at 50 °C ±0.40 °C at 420 °C ±0.60 °C at 660 °C
Radial uniformity Difference between borings	±0.01 °C	±0.01 °C at 50 °C ±0.05 °C at 420 °C ±0.08 °C at 660 °C
Loading effect with internal reference sensor With 4 pcs 6 mm sensors	± 0.08 °C	±0.02 °C at 50 °C ±0.08 °C at 420 °C ±0.15 °C at 660 °C
Loading effect with external 6 mm reference sensor With 3 pcs 6 mm sensors	±0.005 °C	±0.01 °C at 50 °C ±0.02 °C at 420 °C ±0.03 °C at 660 °C
Hysteresis	±0.03 °C	±0.15 °C
Display resolution	0.001 °C /°F / K	0.001 °C /°F / K
Immersion depth	150 mm (5.9 in)	150 mm (5.9 in)
Insert outer dimensions	30 mm (1.18 in)	24.5 mm (0.96 in)
Heating time	23 to 150 °C: 19 min -30 to 150 °C: 23 min	50 to 660 °C: 15 min
Cooling time	150 to 23 °C: 17 min 23 to -30 °C: 23 min 150 to -30 °C: 37 min	660 to 50 °C: 35 min 660 to 100 °C: 25 min
Stabilization time [3]	5 to 10 min	10 min

1) Includes 1-year uncertainty in typical use
2) 30 minutes stability (2 sigma) after the device has reached the setpoint and has stabilized
3) Typical time to stability

Specifications valid at temperature range 13...33 °C if not otherwise specified.

Figure E.1 Beamex MC6-T Multifunction Temperature Calibrator and Communicator data sheet. Beamex Oy Ab, Print: Forsberg, 202021. Reproduced with permission from Beamex.

E.2 Temperature Dry-Block (Handheld)

Fluke Calibration 9102S Handheld Dry-Well Data Sheet

Specifications	9100S	9102S
Range	35°C to 375°C (95°F to 707°F)	−10°C to 122°C (14°F to 252°F) at 23°C ambient
Accuracy	±0.25°C at 50°C; ±0.25°C at 100°C; ±0.5°C at 375°C	±0.25°C
Stability	±0.07°C at 50°C; ±0.1°C at 100°C; ±0.3°C at 375°C	±0.05°C
Well-to-well uniformity	±0.2°C with sensors of similar size at equal depths within wells	
Heating times	35°C to 375°C: 9.5 minutes	Ambient to 100°C: 10 minutes
Stabilization	5 minutes	7 minutes
Cooling times	375°C to 100°C: 14 minutes	Ambient to 0°C: 10 minutes
Well depth	102 mm (4 in); 1.6 mm (1/16 in) hole is 89 mm (3.5 in) deep	102 mm (4 in)
Removable inserts	N/A	Available in sizes from 1.6 mm (1/16 in) to 11.1 mm (7/16 in) [6.4 mm (1/4 in) and 4.8 mm (3/16 in) included]
Power	115 V AC (±10%), 55–65 Hz, 1.5 A or 230 V AC (±10%), 0.8 A, 45–55 Hz, 175 W	94-234 V AC (±10%), 50/60 Hz, 60 W; or 12 V DC
Size (H x W x D)	57 x 125 x 150 mm (2.25 x 4.9 x 6.1 in)	99 x 140 x 175 mm (3.9 x 5.5 x 6.9 in)
Weight	1 kg (2 lb 3 oz)	1.8 kg (4 lb)
Computer interface	RS-232 included	
NIST-traceable calibration	Data at 50°C, 100°C, 150°C, 200°C, 250°C, 300°C and 375°C	Data at −10°C, 0°C, 25 °C, 50°C, 75 °C,100°C and 122°C

Figure E.2 Fluke Calibration 9102S Handheld Dry-Well data sheet. Reproduced with permission from Fluke Corporation.

E.3 Pneumatic Hand Pump and Test Gauge

Fluke 700PTP-1 Pneumatic Test Pump

The Fluke-700PTP-1 pneumatic test pump is capable of supplying pressures up to 600 psi, 40 bar.

For use with 700 series pressure modules or test gauges. Can be directly connected to the reference port consolidating two separate pieces when performing pressure calibration and testing. Can also be used with 717 pressure calibrators. Ideal for calibrating pressure transmitters, pressure switches and pressure gauges up to 600 PSI/40 bar.

Figure E.3 Fluke 700PTP-1 pneumatic test pump. Reproduced with permission from Fluke Corporation.

Fluke 700G Series Precision Pressure Test Gauges

Model	Range	Resolution	Accuracy	Burst pressure
Fluke-700G01	-10 to +10 inH_2O -20 mbar to 20 mbar	0.001 inH_2O 0.001 mbar	Positive pressure ± 0.1 % of range, Vacuum ± 0.1 % of range	3 psi 210 mbar
Fluke-700G02	-1 to +1 psi -70 mbar to 70 mbar	0.0001 psi 0.001 mbar		5 psi 350 mbar
Fluke-700G04	-14 psi to 15 psi -.97 bar to 1 bar	0.001 psi 0.0001 bar	Positive pressure ± 0.05 % FS Vacuum ± 0.1 % FS	60 psi 4 bar
Fluke-700G05	-14 psi to 30 psi -.97 bar to 2 bar	0.001 psi 0.0001 bar		120 psi 8 bar
Fluke-700G06	-12 psi to 100 psi -.83 bar to 6.9 bar	0.01 psi 0.0001 bar		400 psi 26 bar
Fluke-700G27	-12 psi to 300 psi -.83 bar to 20 bar	0.01 psi 0.001 bar		1200 psi 80 bar
Fluke-700G07	-12 psi to 500 psi -.83 bar to 34 bar	0.01 psi 0.001 bar		2000 psi 138 bar
Fluke-700G08	-14 psi to 1000 psi -.97 bar to 69 bar	0.1 psi 0.001 bar		4000 psi 266 bar
Fluke-700G10	-14 to 2000 psi -.97 bar to 140 bar	0.1 psi 0.01 bar		8000 psi 550 bar
Fluke-700G29	-14 psi to 3000 psi -.97 bar to 200 bar	0.1 psi 0.01 bar		10000 psi 690 bar
Fluke-700G30	-14 psi to 5000 psi -.97 bar to 340 bar	0.1 psi 0.01 bar		10000 psi 690 bar
Fluke-700G31	-14 psi to 10000 psi -.97 bar to 690 bar	1 psi 0.01 bar		20000 psi 1035 bar
Fluke-700GA4	0 to 15 psia 0 to 1 bar absolute	0.001 psi 0.0001 bar	± 0.05 % of range	60 psi 4 bar
Fluke-700GA5	0 to 30 psia 0 to 2 bar absolute	0.001 psi 0.0001 bar		120 psi 8 bar
Fluke-700GA6	0 to 100 psia 0 to 6.9 bar absolute	0.01 psi 0.0001 bar		400 psi 27 bar
Fluke-700GA27	0 to 300 psia 0 to 20 bar absolute	0.01 psi 0.001 bar		1200 psi 80 bar
Fluke-700RG05	-14 to +30 psi -0.97 to +2 bar	0.001 psi 0.0001 bar	Positive pressure, ± 0.04 % of reading +.01 % of range, Vacuum ± 0.05 % of range	90 psi 6 bar
Fluke-700RG06	-12 to 100 psi -0.83 to 6.9 bar	0.01 psi 0.0001 bar	Positive pressure, ± 0.04 % of reading +.01 % of range, Vacuum ± 0.1 % of range	400 psi 27 bar
Fluke-700RG07	-12 to 500 psi -0.83 to 34	0.01 psi 0.001 bar		2000 psi 138 bar
Fluke-700RG08	-14 to 1000 psi -0.97 to 69 bar	0.1 psi 0.001 bar		4000 psi 275 bar
Fluke-700RG29	-14 to 3000 psi -0.97 to 200 bar	0.1 psi 0.01 bar		10000 psi 690 bar
Fluke-700RG30	-14 to 5000 psi -0.97 to 340 bar	0.1 psi 0.01 bar		15000 psi 1035 bar
Fluke-700RG31	-14 to 10000 psi -0.97 to 690 bar	1 psi 0.01 bar		20000 psi 1380 bar

700R Ranges: Temperature Compensation 0 °C to 50 °C (32 °F to 122 °F) to rated accuracy.
For temperatures from -10 °C to 0 °C and 50 °C to 55 °C, add .005 % FS/°C

Media compatibility

- 10 inH_2O, 1, 15, 30 psi: any clean dry non-corrosive gas
- 100, 300, 500, 1000 psi: any liquids or gases compatible with 316 stainless steel and above 1000 psi: any non-flammable, non-toxic, non-explosive, non-oxidizing liquid or gas compatible with 316 stainless steel.

Figure E.4 Fluke 700G Series Precision Pressure Test Gauges data sheet. Reproduced with permission from Fluke Corporation.

E.4 Pressure Calibrator (Full Function)

Fluke 729 Pro Automatic Pressure Calibrator

(a)

	729Pro 7M	729Pro 4M
Key parameters		
Pressure control range	(-14 to 1000) psi (-95 to 7000) kPa (-0.95 to 70) bar	(-14 to 600) psi (-95 to 4000) kPa (-0.95 to 40) bar
Pressure type	Gauge/absolute (Depends on the type of pressure module)	
Control stability	0.005% of FS, minimum	
Built-in replaceable pressure control modules	Yes, 6 modules	Yes, 4 modules
External pressure measurement modules	750 Series pressure modules, 50 modules ranging from 0 to 1 in H20, 2.5 mbar, 250 Pa to 10,000 psi, 70 bar, 7 MPa	
mA sourcing/measurement	Yes	
Voltage measurement	Yes	
Loop power	24 V dc	
One-key calibration	Task templates make documented automatic calibrations easy.	
Pressure switch testing	Yes	
Pressure leakage testing	Yes	
HART	Yes	
Autostep / auto ramp	Yes (with customizable speed, mode and style)	
Log function	Yes	
Software	DPCTrack2 Calibration and Asset Management Software	
One-key pressure-relief	Yes (muffler)	
Battery	Rechargeable lithium battery with capacity indication (charged with the instrument or separately)	
Language	English	

Pressure module selection table

Pressure module	Pressure type	Range	Resolution [3]	Overall uncertainty	729Pro 4M	729Pro 7M
FLK-PMM-200K [1]	Gauge	(-15 to 30) psi (-100 to 200) kPa (-1 to 2) bar	0.0001 psi 0.001 kPa 0.00001 bar	0.02 % FS / year, maximum	•	•
FLK-PMM-200KA[2]	Absolute	(0 to 30) psi (0 to 200) kPa (0 to 2) bar	0.0001 psi 0.001 kPa 0.00001 bar	0.05 % FS / year, maximum	•	•
FLK-PMM-2000K [1]	Gauge	(-14 to 300) psi (-95 to 2000) kPa (-0.95 to 20) bar	0.001 psi 0.01 kPa 0.0001 bar	0.02 % FS / year, maximum	•	•
FLK-PMM-4000K [1]	Gauge	(-14 to 600) psi (-95 to 4000) kPa (-0.95 to 40) bar	0.01 psi 0.01 kPa 0.0001 bar	0.02 % FS / year, maximum	•	•
FLK-PMM-7000K [1]	Gauge	(-14 to 1000) psi (-95 to 7000) kPa (-0.95 to 70) bar	0.01 psi 0.01 kPa 0.0001 bar	0.02 % FS / year, maximum		•
FLK-PMM-7000KA[2]	Absolute	(0 to 1000) psi (0 to 7000) kPa (0 to 70) bar	0.01 psi 0.01 kPa 0.0001 bar	0.02 % FS / year, maximum		•

[1] Measurement reference
[2] Absolute reference
[3] Dry air or non-corrosive gases only / 15 °C to 35 °C

Figure E.5a Fluke 729 Pro specifications. Reproduced with permission from Fluke Corporation.

(b)

Specifications

	729Pro 7M	729Pro 4M
Pressure specifications (See User Manual for detail specifications)		
1-year specifications	0.02% of FS	
Control specifications	0.005% of FS, minimum	
Electrical specifications (*All specifications are valid to 110 % of range, except 24 mA sourcing and simulating which are valid to 100 % of range)		
mA measure, source and simulate	0 mA to 24 mA (resolution of 1 μA)	
Volts dc measurement	0 V dc to 30 V dc (resolution of 1 mV)	
Accuracy	0.01% FS ± 2 LSD (23 °C ± 5 °C)	
Stability	20 ppm FS/°C (−10 °C to +18 °C and 28 °C to 50 °C)	
Loop compliance voltage	24 V dc @ 20 mA	
mA simulation working voltage	12 V dc to 30 V dc	
Temperature measurement, 100 Ω Pt (385) RTD	−50 °C to +150 °C	
Temperature resolution	0.01 °C	
Temperature accuracy	±0.1 °C ± 0.25 °C combined uncertainty when using 720 RTD probe (optional accessory)	
Drive capability (mA Source)	1200 Ω (without HART resistor), 950 Ω (with internal HART resistor)	
Mechanical specifications		
Dimensions (H x W x L)	290 mm x 215 mm x 100 mm	
Weight	4.9kg	
Environmental specifications		
Operating temperature	−10 °C to 50 °C	
Pressure control	0 °C to 50 °C	
Battery Charging	0 °C to 40 °C	
Storage temperature	−20 °C to 60 °C	
Operating altitude	<3000 m	
Storage altitude	<13000 m	
Operating humidity	Non-condensing (<10 °C) ≤90 % RH (10 °C to 30 °C) ≤75 % RH (30 °C to 40 °C) ≤45 % RH (40 °C to 50 °C)	
Safety		
General	IEC 61010-1, pollution degree 2	
Lithium battery	IEC 62133, UN 38.3; 14.4 V, 6.7Ah, 97Wh, 4ICR19/66-2 (4s2p) Charge input 19.5 V, 1.6 A	

Figure E.5b Fluke 729 Pro specifications. Reproduced with permission from Fluke Corporation.

Appendix F

Unit Conversions

F.1 Pressure Conversions

Refer to the *ISA Handbook of Measurement Equations and Tables* [1] for more conversion factors.
psi = in. $H_2O \times (0.0361273)$ psi = in. Hg $\times$ (0.491154)
psi = mm $H_2O \times (0.00142233)$ psi = mm Hg $\times$ (0.0193368) psi = kg/cm$^2 \times (14.223)$
psi = bar $\times$ (14.50377)
psi = mbar $\times (0.014503 \times 10^{-2})$ psi = Pa $\times (1.4503 \times 10^{-4})$
psi = kPa $\times$ (0.1450377)
psi = millitorr $\times (1.93368 \times 10^{-5})$

F.2 Temperature Conversions

$$
\begin{aligned}
°\text{F} &= \frac{9}{5}\,°\text{C} + 32 \\
R &= °\text{F} + 459.67 \\
K &= °\text{C} + 273.15
\end{aligned}
$$

Reference

1 Strothman, Jim, ed., *ISA Handbook of Measurement Equations and Tables*, 2nd ed. Research Triangle Park, NC: International Society of Automation (ISA), 2006.

Appendix G

Answers to Review Questions

Chapter 1

1.1 A pressure transmitter's base accuracy is given:
Accuracy = ±0.5% of the *full range*
URL= 200 psig
LRL = 0 psig
a) What is the error when the reading is 50 psig?
b) What is the error when the reading is 100 psig?
Answer: Because the accuracy is given as a percentage of the range, the amount of error in psig will be the same for all readings:
Range = URL LRL
(200 psig – 0 psig) × ±0.005 = 1 psig
a) *At 50 psig: error = ±1 psig*
b) *At 100 psig: error = ±1 psig*

1.2 Explain the difference between the terms *accuracy* and *precision*.
Answer:
Precision = How repeatable can the device measurement be?
Precision indicates how close together a device's measurement data is when the data is recorded under the same conditions by the same person with the same equipment.
Accuracy = How close to "right" can the device measurement be?
Accuracy is the degree to which an indicated value matches the actual value of the measured variable.

1.3 The term *tolerance* means "how close is close enough" or "how good does it need to be?" Is it better to express tolerance in percentage of reading or in actual units of measure?
Answer: Using actual units of measure is the best method to prevent confusion. For example, the tolerance of a pressure measurement might be given as ±0.5 psig.

1.4 A temperature transmitter is capable of measuring –100 °F to 100 °F, but it is configured to measure 0 °F to 50 °F. Calculate the values for:

Answer:

URL 100 °F
LRL –100 °F
URV 50 °F
LRV 0 °F
Range 100 °F – (–100 °F) = 200 °F
Span 50 °F – 0 °F = 50 °F

1.5 A temperature transmitter with a configuration range of 0 °F to 100 °F is found to have the outputs shown in this table:

Input	Expected Output	Actual Output
0 °F	4.0 mA	4.0 mA
25 °F	8.0 mA	8.1 mA
50 °F	12.0 mA	12.2 mA
75 °F	16.0 mA	16.3 mA
100 °F	20.0 mA	20.4 mA

This is an example of ____________.
a) Zero error
b) Span error
c) Both errors combined
d) Hysteresis
Answer: b)
The error is proportional to the input, and there is no offset at the zero value.

1.6 A temperature gauge with a range of 0 °F to 100 °F is found to have these readings as the temperature is increased and then decreased:

Input	Reading	Error
0 °F	0 °F	0
25 °F	25.1 °F	+0.1
50 °F	50.3 °F	+0.3
75 °F	75.1 °F	+0.1
100 °F	100 °F	0
75 °F	74.9 °F	–0.1
50 °F	49.7 °F	–0.3
25 °F	24.9 °F	–0.1
0 °F	0 °F	0

This is an example of __________.

a) Zero error
b) Span error
c) Both errors combined
d) Hysteresis

Answer: d)

The error differs depending on whether the input is increasing or decreasing. This type of error difference is often found in mechanical devices, such as gauges, but it can also be found in electronics.

1.7 ISA-RP105.00.01-2017 says the recommended best practice for performing calibrations is:

a) Calibrate just the instrument (device calibration).
b) Calibrate the entire instrument loop, including the instrument to the control system input, to the human-machine interface (HMI; loop calibration).

Answer: a)

ISA-RP105.00.07-2017 says it is best to calibrate the entire loop when practical. This process checks the function of the instrumentation and the control system.

1.8 Which of the following best describes the term *traceability* of reference standards used for calibration?

a) The ability to calibrate a device using any available reference standard
b) The capability of a field calibrator to store historical calibration data
c) The ability to track the calibrator's accuracy back to a known standard following a clear, documented path of calibrations
d) The use of manufacturer specifications to determine the total probable error of a measurement

Answer: c)

Definition: Traceability is the property of a measurement result whereby the result can be related to a reference through a documented, unbroken chain of calibrations, each contributing to the measurement uncertainty.

1.9 Regarding traceability, the calibration technician should do one or more of the following. Select **all** that apply.

a) Ensure the test equipment is within its calibration interval before using it to perform calibrations.
b) Send test equipment to a certified lab before every use.
c) Record the equipment's unique identifier on the calibration record sheet.
d) Perform bench calibrations on every instrument.

Answer: a) and c)

CCST Level III

1.10 A pressure transmitter's specifications include:

Range = 0 psig to 100 psig
Base accuracy: ±1% of range
Temperature effects: ± 0.5% of range

a) Use the RSS equation to calculate the TPE for the device.
b) Calculate the WCE for the device.

Answer:

$Base\ accuracy = 100\text{ psig} \times 1\% = \pm 1\text{ psig}$

$Temperature\ effect = 100\text{ psig} \times 0.5\% = \pm 0.5\text{ psig}$

a) $TPE = \sqrt{1^2 + 0.5^2} = \pm 1.1\text{ psig}$
b) $WCE = 1\text{ psig} + 0.5\text{ psig} = \pm 1.5\text{ psig}$

Chapter 2

2.1 Which of the following is the primary reason for implementing a calibration program?

a) To ensure all instruments are replaced on a regular schedule
b) To detect and correct instrument errors to ensure measurement accuracy to the required tolerance
c) To reduce the need for preventive maintenance
d) To allow operators to make manual adjustments based on plant conditions

Answer: b)

A calibration program ensures that instruments continue to provide accurate measurements over time.

2.2 What is the first step in establishing a calibration program?

a) Set the calibration frequency for each loop.
b) Set the required tolerance for each loop.
c) Establish a tagging system and create a list of loops or devices that require calibration.
d) Write calibration procedures.

Answer: c)

2.3 If loops associated with regulated emissions monitoring must be calibrated on a specific schedule, then what is a recommended method of classifying them?

Think of some examples in your work where criticality classification could be applied.
Answer: Assign a criticality classification

2.4 Why is it important to assign a criticality classification to instruments in a calibration program?
a) To determine which instruments require protective enclosures
b) To prioritize calibration frequency, capture regulatory requirements, or identify maintenance requirements for each loop
c) To ensure that all instruments are calibrated at the same interval, regardless of function or importance
d) To classify instruments by manufacturer for easier spare parts management
Answer: b)

2.5 A custody transfer contract specifies that the product flow measurement must be accurate to ±2% of the meter range. This information would be used to specify what about the loop calibration?
a) Loop tolerance
b) Meter span, upper range value (URV), lower range value (LRV)
c) Meter accuracy
d) Loop confirmation interval
Answer: a)

2.6 If the meter instrument range in the previous question is 0–1,000 gal./hr., what is the loop tolerance in units of measure?
Answer: 1,000 × 0.02 = ±20 gal./hr.

2.7 What is the purpose of establishing a calibration confirmation interval for a loop?
a) To match the instrument's warranty period
b) To determine how frequently a sensor must be replaced
c) To verify that the instrument continues to perform within tolerance between calibrations
d) To allow operators to bypass alarms during normal operation
Answer: c)

2.8 Calibration confirmation intervals are established for each loop to define how often the loop must be calibrated. How would a reliability-centered maintenance program modify the confirmation intervals?
Answer: If repeated calibrations show that a loop is within tolerance, then the time between calibrations might be increased. If the loop is frequently out of tolerance, then the time between calibrations might be shortened.

2.9 Which document provides step-by-step instructions, acceptance criteria, and equipment required to calibrate a loop?
Answer: The calibration procedure

2.10 Which document is used to record the results of a calibration, such as the as-found and as-left readings?
Answer: The calibration service record (CSR)

Chapter 3

Matching exercise: Choose the correct document(s) associated with each item of information in this list. There might be more than one answer for each item. Some of these are covered in Chapter 2.

Which output device does the instrument control?	1, 2
Is the temperature element installed in a thermowell?	1
What steps should be followed when calibrating a loop?	5
Where should the results be recorded?	6
What is the configured range of a device (lower range value [LRV], upper range value [URV])?	3, 5
What is the maximum range of the device (lower range limit [LRL], upper range limit [URL])?	4
What are the base accuracy, stability, and other effects on uncertainty?	4
What is the allowable loop tolerance?	5

Documentation:

1) Piping and instrumentation diagram (P&ID)
2) Loop diagram
3) Instrument specification sheet
4) Manufacturer's data sheets
5) Calibration procedure
6) Calibration service record

Chapter 4

4.1 What is one big difference between bench calibration equipment and field calibration equipment?

a) Field equipment is typically battery powered and portable.
b) Bench equipment must be certified more often.
c) Field calibration equipment is typically more accurate than bench equipment.
d) Bench equipment is used for mechanical instruments; field calibration equipment is used for electrical instruments.

Answer: a)

4.2 What does the TUR represent in a calibration?

a) The ratio of the calibration equipment's TPE to its base accuracy
b) The ratio of the loop tolerance to the TPE of the calibration equipment
c) The time until the calibration equipment must be recertified
d) The ratio of the standard's accuracy to the ambient temperature

Answer: b)

4.3 What is the recommended minimum TUR for most circumstances?

Answer: $TUR \geq 4$

4.4 Calculate the TUR for this calibration at the full scale of 200 psig:

- Loop tolerance = ±1 psig
- Calibrator TPE = ± 0.5 psig

Is this TUR acceptable?

Answer:

Upper specification limit (USL) = 201

Lower specification limit (LSL) = 199

$$TUR = \frac{201 - 199}{2 \times 0.5} = \frac{2}{1} = 2$$

TUR<4, so it is not acceptable.

4.5 What is the purpose of having calibration equipment calibrated by a certified lab?

a) To ensure the equipment is new and unused
b) To reduce the need for documenting measurements
c) To ensure the measurement is traceable to national and international standards through an unbroken chain of calibrations
d) To enable faster measurements

Answer: c)

4.6 How can a CST know the calibration equipment is within its certification interval?

Answer: The calibration sticker attached to the device with the next due date of calibration

4.7 Why is it important to know if a device must be adjusted for local conditions?

a) To prevent electronic interference
b) So the equipment can be used without a calibration procedure
c) Because temperature, atmospheric pressure, and local gravity can affect measurements

Answer: c)

Chapter 5

5.1 Two common temperature elements are RTDs and thermocouples. Which device outputs voltage related to temperature, and which changes resistance with temperature?

Answer:

Thermocouple output voltage is related to the *temperature. The voltage difference between the reference junction and the measuring junction increases as the temperature difference between them increases.*

RTDs change resistance with the *temperature. The resistance increases as the temperature increases.*

5.2 Why is it important to know if a temperature element is installed in a thermowell?

a) Thermowells change the properties of a sensor.
b) Thermowells eliminate the need for calibration.
c) If a sensor is in a thermowell, then it can be removed for calibration without shutting off the process.
d) The thermowell requires special calibration equipment.

Answer: c)

5.3 What is the function of a temperature transmitter?

a) Convert a raw RTD or thermocouple signal to a standard output (such as 4–20 mA) for the control system
b) Control the process temperature based on a set point
c) Protect the sensor from the process fluids
d) Improve the sensor accuracy

Answer: a)

5.4 Which components must be considered for calculating a temperature loop TPE?

Answer:

- *Sensor (RTD, thermocouple)*
- *Transmitter*
- *Controller input card*

5.5 Why might a procedure call for using an external reference probe in a dry-block calibrator instead of using the internal reference probe?

a) The external probe eliminates the need for stabilization time.
b) The external probe does not require power.
c) The built-in reference can only be used for certain types of sensors.
d) The external probe provides a more accurate measurement comparison with the device being tested.

Answer: d)

The internal probe is in the heated block, whereas the external probe is usually in the same well as the device being tested.

5.6 What is the maximum error in degrees Celsius that is expected of an ASTM Grade A RTD at 50 °C? (Use the RTD specification in Appendix C.)

Answer:

$$\text{Grade A} = \pm\lfloor 0.13 + 0.0017|t|\rfloor\,^{\circ}\text{C}$$

$$\text{Error} = \pm 0.22\,^{\circ}\text{C at } 50\,^{\circ}\text{C}$$

5.7 For the loop calibration example in Section 5.5.4.2, calculate the percentage reading error for each input point listed in Figure 5.15.

Answer: Using Equation 5.7

Input (%)	%Reading Error
0	−100.00
25	−1.00
50	−0.88
75	−0.97
100	−1.34

5.8 Regarding the example in Section 5.5.4.4, if the sensor was a thermocouple instead of an RTD, what kind of signal would be simulated to the transmitter input?

Answer: A voltage instead of a resistance

5.9 Which equipment would ***not*** be used when calibrating a temperature loop or temperature sensor?
a) Digital multimeter
b) Dry-block (dry-well)
c) Temperature bath
d) Hand pump
Answer: d)

Chapter 6

6.1 If a gauge pressure device reads 110 psig, what would the absolute pressure reading be if the device were in:
1) Denver, Colorado (atmospheric pressure = 12.1 psia)?
Answer: 110 psig + 12.1 psia = 122.1 psia
2) Houston, Texas (atmospheric pressure = 14.7 psia)?
Answer: 110 psig + 14.7 psia = 124.7 psia

6.2 Using the piping and instrumentation diagram (P&ID; DRG P&ID 15) in Appendix A, identify the components in pressure loop 303. List the tags. What is the type and function of each device?
Answer:
PIT-303, pressure transmitter senses the pressure and converts it to a signal for the controller
PIC-303, pressure indicating controller, probably in a distributed control system (DCS) or programmable logic controller (PLC), reads the output from the transmitter, displays it to the operator, and sends an output to control the valve to maintain the pressure at the set point
PY-303 current-to-pressure (I/P) converter converts the electrical signal from the PLC output to a pressure signal for the valve actuator
PV-303 pressure valves throttle open or close to control the flow

6.3 There would be two calibration procedures for the loop in Question 6.2. One for the input side and one for the output side of the controller. List the components that would be in each procedure.
Answer:
Input side procedure:
- *PIT-303 (attach a pressure source to the transmitter)*
- *PIC-303 (read the pressure value on the display)*

Output side procedure
- *PIC-303 (manually change the output value)*
- *PY-303*
- *PV-303 (confirm the valve position)*

6.4 Does the transmitter in Question 6.2 have both a block valve and a bleed valve to allow attaching the pressure source without removing the transmitter from service?
Answer: No. The piping and instrumentation diagram (P&ID) only shows a block valve. The block valve could be closed to remove the transmitter for calibration.

6.5 Summarize the main steps in calibrating the input loop identified in Question 6.3.
Answer:
1) *Use a pressure source, such as a hand pump, to apply pressure to the transmitter.*
2) *Read the applied pressure on a calibrated device. Record the reading in the "applied input" column of the calibration service record (CSR).*
3) *Read the pressure shown on the operator display. Record the reading in the as-found column of the CSR.*
4) *Calculate the error for each reading.*
5) *Determine if the readings are acceptable (within tolerance).*

6.6 If the loop tolerance is ±1 psig, what is the highest desirable value for the TPE of the calibrated device used to read the applied pressure?
Answer: ±0.25 psig (It must be at least 4 times better than the loop tolerance.)

6.7 A pressure gauge on a nozzle near the bottom of a tank reads 5.0 psig. The tank is open to the atmosphere, so there is no blanket gas pressure. The specific gravity of the fluid in the tank is 0.8. Calculate the height of the fluid in the tank.

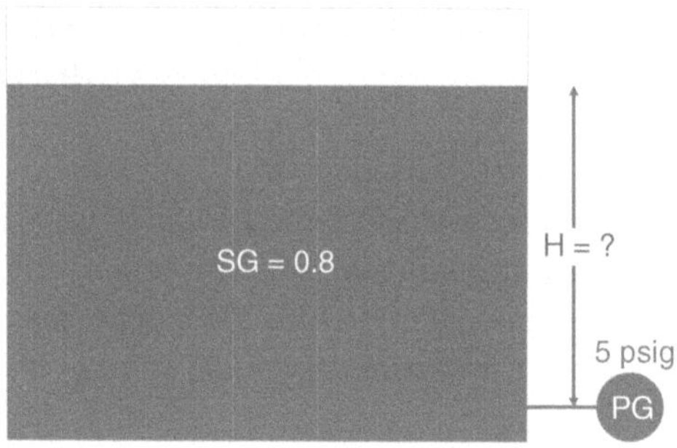

Answer: 14.4 feet above the nozzle:

$$\frac{5 \text{ psig}}{0.8} \times \frac{2.31 \text{ ft. } H_2O}{\text{psig}} = 14.4 \text{ ft.}$$

6.8 What test equipment would be needed to calibrate an electronic pressure transmitter?
a) Pressure standard, current standard
b) Current standard, communicator
c) Pressure standard, analog voltmeter
d) Pressure standard
Answer: a)

6.9 If an elephant weighing 12,000 lb. stands on one foot that has an area of 200 in.^{2}, how much pressure is under that foot?
Answer:

$$P = \frac{\text{force}}{\text{area}} = \frac{1 \text{ elephant}}{\text{elephant foot}} = \frac{12{,}000 \text{ lb.}}{200 \text{ in.}^2} = 60 \text{ psig}$$

6.10 If a human weighing 150 lb. stands on one foot that has an area of 30 in.^{2}, how much pressure is under that foot?
Answer:

$$P = \frac{\text{force}}{\text{area}} = \frac{\text{1 human}}{\text{human foot}} = \frac{\text{150 lb.}}{\text{30 in.}^2} = 5 \text{ psig}$$

6.11 If the elephant stands on the human's foot, then what?
Answer: Ouch

Chapter 7

7.1 Hydrostatic pressure level measurement is an example of:
a) Indirect measurement
b) Direct measurement
Answer: a)
Pressure is measured to infer the level.

7.2 If hydrostatic pressure is used for level measurement in a pressurized vessel, then why must the vessel pressure be measured in addition to the hydrostatic pressure at the bottom?
a) To calculate the volume of fluid in the tank from the level
b) To compensate for the pressure exerted by the gas vapor space above the liquid
c) To correct for sensor drift over time
d) To determine the density of the process fluid
Answer: b)

7.3 Why is it important to consider the effect of the temperature on the capillary fill fluid in a remotely sealed hydrostatic level transmitter?
a) High temperatures cause the capillary fluid to dissolve into the process fluid.
b) Fill fluid evaporates at high temperatures.
c) It is not important.
d) Temperature changes can expand or contract the fill fluid, affecting the pressure reading on the DP transmitter and introducing error.
Answer: d)

7.4 How can temperature affect the level of water in a vessel?
a) Temperature has no effect on the level of water in a vessel.
b) As the temperature increases, water expands, raising the actual level in the vessel.
c) As the temperature increases, water density increases and causes the level to drop.
Answer: b)

7.5 Why is a manual tape gauge often used to calibrate a level loop?
a) It directly measures the pressure in the vessel.
b) It is the only method approved by international standards.

c) It directly measures the actual level in the vessel for comparison to the loop measurement.
d) It automatically logs data for historical records.
Answer: c)

CCST Level III

7.6 A radar level transmitter has these specifications:

- Reference accuracy: ±0.12 in. (at 77 °F)
- Repeatability: ±0.04 in.
- Ambient temperature effect: ±0.0044 in. per degree Fahrenheit from reference temperature

a) What is the TPE of the radar device at an ambient temperature of 50 °F?
Answer:
Ambient temperature effect at 50° F:

$$(77\,°F - 50\,°F) \times \pm 0.0044 \frac{\text{in.}}{°\text{F}} = \pm 0.12\text{ in.}$$

Calculate TPE transmitter:

$$TPE = \pm\sqrt{0.12^2 + 0.04^2 + 0.12^2} = \pm 0.17\ in$$

b) The transmitter is set to measure a range of 0 in. to 120 in. The transmitter 4–20 mA output is connected to a PLC input that has an uncertainty of ±0.1% of the range. What is the TPE of the temperature measurement loop?
Answer:
Calculate loop TPE:

$$\text{Input card error: } \pm 0.1\% \times 120\text{ in.} = 0.12\text{ in.}$$

$$\text{loop TPE} = \pm\sqrt{0.17^2 + 0.12^2} = 0.21\text{ in.}$$

Chapter 8

8.1 Why is in situ calibration often preferred over bench calibration for a flowmeter?
a) It verifies meter accuracy under actual process conditions with the actual process fluid.
b) Field reference standards are cheaper and easier to use.
c) It can be performed without a formal procedure.
d) It eliminates the need to involve maintenance personnel.
Answer: a)

8.2 After proving a liquid meter and calculating a meter factor, why might a meter be adjusted or replaced instead of just applying the new meter factor?
a) Meter factors can only be used once.
b) Adjusting the meter is easier.
c) Meter factors are not allowed in regulated applications.
d) A significant change in meter factor from the previous proving may indicate wear, damage, or malfunction.
Answer: d)

8.3 To calibrate an orifice meter in gas service, the typical method would be to:
a) Inspect the primary devices (orifice plate and holder) and calibrate the secondary devices (DP, pressure, and temperature sensors).
b) Use meter diagnostics.
c) Flow into a proving tank.
Answer: a)

8.4 What is the main purpose of proving a liquid flowmeter?
a) To verify the pressure rating of the meter under maximum flow
b) To determine a correction factor (meter factor) that adjusts the meter's output to match a known reference value
c) To clean the internals of the meter using calibration fluid
d) To verify that the meter's tag number matches the drawings
Answer: b)

8.5 How is meter proving accomplished for liquid meters?
a) By comparing the flowmeter's indicated volume to a known reference volume at actual flowing conditions
b) By disconnecting the meter and testing it on a bench with water
c) By averaging the readings from multiple flowmeters in the same line
Answer: a)

Chapter 9

9.1 Refer to the piping and instrumentation diagram (P&ID) in Appendix A. What are the final control elements in temperature loop T-300?
Answer:
TY-300 – I/P converter
TV-300 – control valve

9.2 What is the main objective of calibrating a final control element, such as a control valve?
a) Ensure the valve is fully closed during shutdown
b) Verify that the valve position accurately responds to the control signal over its full range of operation
c) Reduce air consumption in pneumatic actuators
d) Confirm that the automatic controller is working correctly
Answer: b)

9.3 What is the role of a final control element in a control system?
a) Measure process variables like pressure and temperature
b) Calculate set points for automatic controllers
c) Convert control signals into physical action that directly affects the process
d) Transmit data to the control room for monitoring
Answer: c)

9.4 Summarize the basic steps to perform a loop calibration on a control valve.
Answer:
1) *Put the loop controller in manual.*
2) *Change the controller output to different values (e.g., 0%, 50%, and 100%).*
3) *Verify the valve is in the commanded position.*
4) *Record the results.*

Chapter 10

10.1 What is the primary objective when calibrating analytical process instruments like pH meters or gas analyzers?
a) Ensure the sensor stays clean during the process operation
b) Verify the instrument's output accurately corresponds to a known reference standard
c) Increase the sample frequency for better process control.
Answer: b)

10.2 What kind of reference standard is typically used for calibrating analytical instrumentation such as pH meters, conductivity sensors, or gas analyzers?
a) Pressure gauges and temperature probes
b) Electrical signal generators
c) Certified buffer solutions or calibration gases with known chemical composition
d) Flowmeters and orifice plates
e) Gauge tapes
Answer: c)

10.3 Why are zero and span checks, or two-point checks, important when calibrating an analytical instrument?
a) They verify that the instrument responds correctly at the low and high ends of the measurement range.
b) They detect moisture buildup on the sensor.
c) They extend the calibration interval.
d) They determine the loop tolerance.
Answer: a)

10.4 Should loop calibration be used on analytical instruments when possible?
Answer: Yes

Chapter 11

11.1 Why is it important to regularly calibrate fixed gas detectors in industrial environments?
a) To reduce the instrument's power consumption
b) To ensure the detector responds accurately to hazardous gas concentrations and provides reliable alarms for people's safety
c) To allow operators to disable alarms for maintenance
d) To adjust the control system's temperature compensation
Answer: b)

11.2 Summarize the steps typically involved in calibrating a gas detector.

Answer:

1) *Make sure the sensor is in clean, fresh air, and verify the zero reading.*
2) *Apply the test gas with a known concentration of the gas of interest.*
3) *Verify that the detector reads correctly.*
4) *Record the result.*
5) *Remove the test gas and verify that the sensor returns to zero.*

Bibliography

Note: The dates and editions for standards and other references in this chapter are current as of publication. The reader is encouraged to check for later revisions or editions.

Industry Standards, Practices, and Guidelines

Advancing Standards Transforming Markets (ASTM). ASTM E1137/E1137M-08(2020), *Standard Specification for Industrial Platinum Resistance Thermometers*. West Conshohocken, PA: ASTM International, reapproved 2020.

American Petroleum Institute (API). *Manual of Petroleum Measurement Standards*. Chapter 1 – "Vocabulary". 2nd ed. Washington, DC: API, 1994.

American Petroleum Institute (API). *Manual of Petroleum Measurement Standards*. Chapter 14.3, parts 1 through 4 – "Manual of Petroleum Measurement Standards, Orifice Metering of Natural Gas and Other Related Hydrocarbon Fluids – Concentric, Square-edged Orifice Meters". 4th ed. Washington, DC: API, 2012.

American Petroleum Institute (API). *Manual of Petroleum Measurement Standard*. Chapter 3.1A – "Standard Practice for the Manual Gauging of Petroleum and Petroleum Products". 3rd ed. Washington, DC: API, 2013.

American Petroleum Institute (API). *Manual of Petroleum Measurement Standards*. Chapter 22 – "Testing Protocol," Section 2 – "Differential Pressure Flow Measurement Devices". 2nd ed. Washington, DC: API, 2017. Reaffirmed, February 2022.

American Petroleum Institute (API). *Manual of Petroleum Measurement Standards*. Chapter 3.1B – "Standard Practice for Level Measurement of Liquid Hydrocarbons in Stationary Tanks by Automatic Tank Gauging". 4th ed. Washington, DC: API, 2021.

American Petroleum Institute (API). *Manual of Petroleum Measurement Standards*. Chapter 21.1 – "Flow Measurement Using Electronic Metering Systems – Electronic Gas Measurement." Washington, DC: API, 2013.

American Society of Mechanical Engineers (ASME). ASME B40.100-2022, *Pressure Gauges and Gauge Attachments*. New York, NY: ASME, 2022.

Instrument Society of America (ISA). ISA-5.4-1991, *Instrument Loop Diagrams*. Research Triangle Park, NC: ISA, 1991.

Instrument Society of America (ISA). ISA-51.1-1979 (R1993), *Process Instrumentation Terminology*. Research Triangle Park, NC: ISA, 1993. This standard has been archived.

International Organization for Standardization (ISO). ISO 6142-1:2015, *Gas analysis – Preparation of Calibration Gas Mixtures – Part 1: Gravimetric Method for Class I Mixtures*. Geneva, Switzerland: ISO, 2015.

International Society of Automation (ISA). ISA TR108.1-2020, *Intelligent Device Management – Part 1: Concepts and Terminology*. Research Triangle Park, NC: ISA, 2020.

International Society of Automation (ISA). ISA-RP105.00.01-2017, *Management of a Calibration Program for Industrial Automation and Control Systems*. Research Triangle Park, NC: ISA, 2017.

International Society of Automation (ISA). ISA 75.13.01-2013, *Method of Evaluating the Performance of Positioners with Analog Input Signals and Pneumatic Output*. Research Triangle Park, NC: ISA, 2013.

International Society of Automation (ISA). ISA-75.25.01-2000 (R2010), *Test Procedure for Control Valve Response Measurement for Step Inputs*. Research Triangle Park, NC: ISA, 2010.

Process Industry Practices (PIP). PIC001, *Piping and Instrumentation Diagram Documentation Criteria*. Technical Revision. Austin, TX: PIP, 2026.

The Instrumentation, Systems, and Automation Society (ISA). ISA-TR20.00.01-2007, *Specification Forms for Process Measurement and Control Instruments – Part 1: General Considerations*. Research Triangle Park, NC: ISA, 2007.

The Instrumentation, Systems, and Automation Society (ISA). ISA-TR91.00.02-2003, *Criticality Classification Guideline for Instrumentation*. Research Triangle Park, NC: ISA, 2003.

The International Electrotechnical Commission (IEC). IEC 60751, *Industrial Platinum Resistance Thermometer Sensors*. Geneva, Switzerland: IEC, 2022.

Vendor References

Beamex. *Ultimate Calibration*. Finland: Beamex Oy Ab, 2012.

Beamex Oy Ab. "Temperature Calibrators." https://www.beamex.com/calibrators/temperature-calibrators/. Accessed April 5, 2025.

Crane. *Flow of Fluid Through Valves, Fittings, and Pipe*. Technical Paper 410. Chicago: Crane Co., 1965.

Emerson. *The Engineer's Guide to Industrial Temperature Measurement*. Shakopee, MN: Emerson, 2013.

Emerson. *Control Valve Handbook*. 5th ed. Marshalltown, IA: Emerson, 2017.

Emerson. *The Engineer's Guide to Overfill Prevention*. Shakopee, MN: Emerson, 2018.

Emerson. *The Engineer's Guide to Level Measurement*. Shakopee, MN: Emerson, 2019.

Emerson. "Calculating Transmitter Calibration Intervals." Technical Note 00840-1900-4801, Rev AA. March 2021.

Emerson Process Management. "Calibrating Conductivity Sensors." Application Data Sheet ADS 43-024. July 2010.

Fluke Calibration. "Fluke Calibration 9100S Handheld Dry-Well." https://www.fluke.com/en-us/product/calibration-tools/temperature-calibrators/fluke-calibration-9100s. Accessed April 5, 2025.

Fluke Corporation. "Understanding and comparing instrument specifications." Application Note. 2011.

Fluke Corporation. *Temperature Calibration Applications and Solutions*. 2015.

Laurila, Heikki. "How to calibrate temperature instruments [Webinar]." Beamex Blog, October 18, 2018. https://blog.beamex.com/how-to-calibrate-temperature-instruments-webinar. Accessed April 5, 2025.

Laurila, Heikki. "Uncertainty components of a temperature calibration using a dry block." Beamex Blog, August 23, 2018. https://blog.beamex.com/uncertainty-components-temperature-calibration-dry-block. Accessed April 5, 2025.

Laurila, Heikki. "Calibration uncertainty and why technicians need to understand it [Webinar]." Beamex Blog, July 27, 2019. https://blog.beamex.com/calibration-uncertainty-and-why-technicians-need-to-understand-it-webinar. Accessed April 5, 2025.

Mettler Toledo Process Analytics. *A guide to online conductivity measurement, theory and practice*. 2016.

MSA: The Safety Company. *Gas Detection Handbook: Key Concepts and Reference Material for Installed Gas Monitoring Systems*. 9th ed. MSA, 2022.

Books and Publications

Freidmann, Paul G. *The Automation, Systems, and Instrumentation Dictionary*. 4th ed. Research Triangle Park, NC: ISA (The Instrumentation, Systems, and Automation Society), 2003.

Grossman, Jeff. "The Electronic Deadweight Tester." *Cal Lab The International Journal of Metrology 16.3* (2009): 36–39.

Meier, Frederick A., and Clifford A. Meier. *Instrumentation and Control Systems Documentation*. 2nd ed. Research Triangle Park, NC: ISA (International Society of Automation), 2011.

Miller, Richard W. *Flow Measurement Engineering Handbook*. 3rd ed. Boston, MA: McGraw-Hill, 1996.

Sands, Nicholas P., and Ian Verhappen, eds. *A Guide to the Automation Body of Knowledge*. 4th ed. Research Triangle Park, NC: ISA (International Society of Automation), 2026.

Index

Wiley The manufacturer's authorized representative according to the EU General Product Safety Regulation is Wiley-VCH GmbH, Boschstr. 12, 69469 Weinheim, Germany, e-mail: Product_Safety@wiley.com.

Printed and bound by CPI Group (UK) Ltd, Croydon, CR0 4YY
28/07/2026
02179784-0001